Advances in Intelligent and Soft Computing 73

Editor-in-Chief: J. Kacprzyk

Advances in Intelligent and Soft Computing

Further volumes of this series can be found on our homepage: springer.com

Vol. 57. M. Kurzynski, M. Wozniak (Eds.)
Computer Recognition Systems 3, 2009
ISBN 978-3-540-93904-7

Vol. 58. J. Mehnen, A. Tiwari, M. Köppen, A. Saad (Eds.)
Applications of Soft Computing, 2009
ISBN 978-3-540-89618-0

Vol. 59. K.A. Cyran, S. Kozielski, J.F. Peters, U. Stańczyk, A. Wakulicz-Deja (Eds.)
Man-Machine Interactions, 2009
ISBN 978-3-642-00562-6

Vol. 60. Z.S. Hippe, J.L. Kulikowski (Eds.)
Human-Computer Systems Interaction, 2009
ISBN 978-3-642-03201-1

Vol. 61. W. Yu, E.N. Sanchez (Eds.)
Advances in Computational Intelligence, 2009
ISBN 978-3-642-03155-7

Vol. 62. B. Cao, T.-F. Li, C.-Y. Zhang (Eds.)
Fuzzy Information and Engineering Volume 2, 2009
ISBN 978-3-642-03663-7

Vol. 63. Á. Herrero, P. Gastaldo, R. Zunino, E. Corchado (Eds.)
Computational Intelligence in Security for Information Systems, 2009
ISBN 978-3-642-04090-0

Vol. 64. E. Tkacz, A. Kapczynski (Eds.)
Internet – Technical Development and Applications, 2009
ISBN 978-3-642-05018-3

Vol. 65. E. Kącki, M. Rudnicki, J. Stempczyńska (Eds.)
Computers in Medical Activity, 2009
ISBN 978-3-642-04461-8

Vol. 66. G.Q. Huang, K.L. Mak, P.G. Maropoulos (Eds.)
Proceedings of the 6th CIRP-Sponsored International Conference on Digital Enterprise Technology, 2009
ISBN 978-3-642-10429-9

Vol. 67. V. Snášel, P.S. Szczepaniak, A. Abraham, J. Kacprzyk (Eds.)
Advances in Intelligent Web Mastering - 2, 2010
ISBN 978-3-642-10686-6

Vol. 68. V.-N. Huynh, Y. Nakamori, J. Lawry, M. Inuiguchi (Eds.)
Integrated Uncertainty Management and Applications, 2010
ISBN 978-3-642-11959-0

Vol. 69. E. Piętka and J. Kawa (Eds.)
Information Technologies in Biomedicine, 2010
ISBN 978-3-642-13104-2

Vol. 70. XXX

Vol. 71. XXX

Vol. 72. J.C. Augusto, J.M. Corchado, P. Novais, C. Analide (Eds.)
Ambient Intelligence, Future Trends - International Symposium on Ambient Intelligence (ISAmI 2010), 2010
ISBN 978-3-642-13267-4

Vol. 73. E. Corchado, P. Novais, C. Analide, J. Sedano (Eds.)
Soft Computing Models in Industrial and Environmental Applications, 5th International Workshop (SOCO 2010), 2010
ISBN 978-3-642-13160-8

Emilio Corchado, Paulo Novais,
Cesar Analide and Javier Sedano (Eds.)

Soft Computing Models in Industrial and Environmental Applications, 5th International Workshop (SOCO 2010)

Editors

Emilio Corchado
Departamento de Informática
y Automática
Facultad de Ciencias
Universidad de Salamanca
Plaza de la Merced S/N
37008, Salamanca
Spain
E-mail: escorchado@usal.es

Cesar Analide
Universidade do Minho
Departamento de Informática
Campus de Gualtar
4710-057 Braga
Portugal
E-mail: analide@di.uminho.pt

Paulo Novais
Universidade do Minho
Departamento de Informática
Campus de Gualtar
4710-057 Braga
Portugal
E-mail: pjon@di.uminho.pt

Javier Sedano
Departamento de Ingeniería
Electromecánica
Universidad de Burgos
Avenida Cantaria S/N
09006 Burgos
E-mail: jsedano@ubu.es

ISBN 978-3-642-13160-8 e-ISBN 978-3-642-13161-5

DOI 10.1007/978-3-642-13161-5

Advances in Intelligent and Soft Computing ISSN 1867-5662

Library of Congress Control Number: 2010927160

Typeset & Cover Design: Scientific Publishing Services Pvt. Ltd., Chennai, India.

Printed on acid-free paper

5 4 3 2 1 0

springer.com

Preface

This volume of Advances in Intelligent and Soft Computing contains accepted papers presented at SOCO 2010 held in the beautiful and historic city of Guimarães, Portugal, June 2010.

The global purpose of SOCO conferences has been to provide a broad and interdisciplinary forum for soft computing and associated paradigms, which are playing increasingly important roles in an important number of industrial and environmental applications fields.

Soft computing represents a collection or set of computational techniques in machine learning, computer science and some engineering disciplines, which investigate, simulate and analyze very complex issues and phenomena. This workshop is mainly focused on its industrial and environmental applications.

SOCO 2010 is the 5th International Workshop on Soft Computing Models in Industrial Applications and provides interesting opportunities to present and discuss the latest theoretical advances and real world applications in this multidisciplinary research field.

This volume presents the papers accepted for the 2010 edition, both for the main event and the Special Sessions. SOCO 2010 Special Sessions are a very useful tool in order to complement the regular program with new or emerging topics of particular interest to the participating community. Special Sessions that emphasize on multi-disciplinary and transversal aspects, as well as cutting-edge topics were especially encouraged and welcome.

SOCO 2010 included a total of 3 Special Sessions: Ensemble Learning and Information Fusion for Industrial Applications; Soft Computing for Service Management; Hybrid Intelligent Systems and Applications.

With a full programme composed of 29 long papers and 2 short papers, these proceedings capture the most innovative results and advances in 2010. Each paper has been reviewed by, at least, three different reviewers, from an international committee composed of 57 members from 15 countries.

The selection of papers was extremely rigorous in order to maintain the high quality of the conference and we would like to thank the members of the Program Committee for their hard work in the reviewing process. This is a crucial process to the creation of a workshop high standard and the SOCO conference would not exist without their help.

SOCO 2010 enjoyed outstanding keynote speeches by distinguished guest speakers: Professor Ajith Abraham, from MIR Labs, Europe, and Professor Magdy Bayoumi , form University of Louisiana at Lafayette, USA).

We would like to thank all the Special Session organizers, contributing authors and the Local Organizing Committee for their hard and highly valuable work. Their work contributed, definitively, to the success of the SOCO 2010 event.

Particular thanks go, as well, to the Workshop main Sponsors: APPIA - Portuguese Association for Artificial Intelligence; AEPIA - Spanish Association for Artificial Intelligence; MIR - Machine Intelligence Research Labs; IEEE Computational Intelligence Society, Portugal Chapter; CCTC – Computer Science and Technology Center and the Department of Informatics of the University of Minho, who jointly contributed in an active and constructive manner to the success of this initiative.

June 2010 The Editors

Emilio Corchado
Paulo Novais
Cesar Analide
Javier Sedano

Organization

Scientific Committee Chairs

Emilio Corchado	University of Salamanca (Spain)
Javier Sedano	University of Burgos (Spain)

Organizing Committee Chairs

Paulo Novais	University of Minho (Portugal)
Cesar Analide	University of Minho (Portugal)

Honorary Co-chair

Ana Madureira	Portuguese IEEE CIS Chapter (Portugal)

Program Committee

Abraham Ajith	Norwegian University of Science and Technology (Norway)
Alberto Freitas	University of Porto (Portugal)
Alvaro Herrero	University of Burgos (Spain)
Ana Almeida	Polytechnic of Porto (Portugal)
António Abelha	University of Minho (Portugal)
Bogdan Gabrys	Bournemouth University (United Kingdom)
Bruno Baruque	University of Burgos (Spain)
Camelia Chira-Babes	Bolyai University (Romany)
Carlos Pereira	Polytechnic of Coimbra (Portugal)
Carlos Redondo Gil	University of León (Spain)
Cesar Analide	University of Minho (Portugal)
David Meehan	Dublin Institute of Technology (Ireland)
Eduardo J. Solteiro Pires	University of Trás-os-Montes e Alto Douro (Portugal)
Emilio Corchado	University of Salamanca (Spain)
Enrique Herrera-Viedma	University of Granada (Spain)
Felix Sanchez	Promatic System (Spain)

Florentino Fernández Riverola	University of Vigo (Spain)
Francisco Herrera	University of Granada (Spain)
Gerald Shaeper	Loughborough University (United Kingdom)
Gregorio Sainz	CARTIF Technological Centre (Spain)
Huiyu Zhou	Queen's University Belfast (United Kingdom)
Javier Sedano	University of Burgos (Spain)
Jorge Lopes	BRISA S.A. (Portugal)
José Alfredo F. Costa	Rio Grande de Norte Federal University (Brazil)
José F. Martínez	INAOE (Mexico)
José Manuel Benítez Sánchez	University of Granada (Spain)
Jose Manuel Molina	Universidad Carlos III de Madrid (Spain)
Jose Villar	University of Oviedo (Spain)
Juan M. Corchado	University of Salamanca (Spain)
Luciano Sanchez Ramos	University of Oviedo (Spain)
Luis Correia	University of Lisboa (Portugal)
Luís Nunes	ISCTE (Portugal)
Luís Paulo Reis	University of Porto (Portugal)
Maciej Grzenda	Warsaw University of Technology (Poland)
Marco Mora	Universidad Católica del Maule (Chile)
Maria João Viamonte	Polytechnic of Porto (Portugal)
Mario Köppen	Kyushu Institute of Technology (Japan)
Michal Wozniak	Wroclaw University of Technology (Poland)
Miroslav Bursa	Czech Technical University (Czech)
Paulo Cortez	University of Minho (Portugal)
Paulo Moura Oliveira	University of Trás-os-Montes e Alto Douro (Portugal)
Paulo Novais	University of Minho (Portugal)
Pedro M. Caballero	CARTIF Technological Centre (Spain)
Petro Gopych	Universal Power Systems USA-Ukraine LLC (Ukraine)
Rosa Basagoiti	Mondragon University (Spain)
Rui Sousa	University of Minho (Portugal)
Sara Silva	INESC ID Lisboa (Portugal)
Sebastian Ventura Soto	University of Córdoba (Spain)
Sergio Saludes	CARTIF Technological Centre (Spain)
Stefano Pizzuti	Energy New technology and Environment Agency (Italy)
Susana Ferreiro Del Río	TEKNIKER (Spain)
Urko Esnaola	Fatronik Tecnalia Fundation (Spain)
Vicente Martin	Universidad Politécnica de Madrid (Spain)
Victor Alves	University of Minho (Portugal)
Wei-Chiang Hong	Oriental Institute of Technology (Taipei)
Yin Hujun	University of Manchester (United Kingdom)
Zita Vale	Polytechnic of Porto (Portugal)

Special Session on Ensemble Learning and Information Fusion Committee

Michal Wozniak	Wroclaw University of Technology (Poland)
Robert Burduk	Wroclaw University of Technology (Poland)
Konrad Jackowski	Wroclaw University of Technology (Poland)
Krzysztof Walkowiak	Wroclaw University of Technology (Poland)
Bruno Baruque	University of Burgos (Spain)
Emilio Corchado	University of Salamanca (Spain)

Special Session on Hybrid Intelligent Systems and Applications Committee

Soumya Banerjee	Birla Institute of Technology (India)
Aboul Ella Hassanien	Cairo University (Egypt)
Mostafa A. El-Hosseini	Kafrelsheikh University (Egypt)

Special Session on Soft Computing for Service Management Committee

Mehmet Aydin	University of Bedfordshire (United Kingdom)
Turkay Dereli	University of Gaziantep (Turkey)

External Reviewers

Arkadiusz Grzybowski	University of Legnica (Poland)
Dragan Simic	University of Novi Sad (Serbia)
Huseyin Seker	DeMonfort University (United Kingdom)
Issam Damaj	American University (Kuwait)
Osman Taylan	King Abdulaziz University (Saudi Arabia)
Stefano Rovetta	University of Genova (Italy)

Local Organization Committee

Ângelo Costa	University of Minho (Portugal)
Cesar Analide	University of Minho (Portugal)
Davide Carneiro	University of Minho (Portugal)
Luís Lima	Polytechnic of Porto (Portugal)
Luís Machado	University of Minho (Portugal)
Paulo Moura Oliveira	University of Trás-os-Montes e Alto Douro (Portugal)
Paulo Novais	University of Minho (Portugal)
Pedro Gomes	University of Minho (Portugal)
Ricardo Costa	Polytechnic of Porto (Portugal)
Sara Fernandes	University of Minho (Portugal)

Contents

Agents and Multiagent Systems

Intelligent Systems

Evolutionary Computing

Energy and Environmental Applications

Hybrid Systems

Applications

Hybrid Intelligent Systems and Applications

A Security Proposal Based on a Real Time Agent to Protect Web Services Against DoS Attack

Cristian Pinzón, Angélica González, Manuel Rubio, and Javier Bajo

Abstract. This paper describes a novel proposal based on a real time agent to detect and block denial of service attacks within web services environments. The real time agent incorporates a classification mechanism based on a Case-Base Reasoning (CBR) model, where the different CBR phases are time bounded. In addition, the reuse phase of the CBR cycle incorporates a mixture of experts to choose a specific technique of classification depending on the feature of the attack and the available time to solve the classification.

Keywords: Multi-agent System, CBR, Web Service, SOAP Message, DoS attacks.

1 Introduction

New security issues as well as new ways of exploiting inherited old security threats can become a serious problem to applications based on web services. One of the threats that is becoming more common within web services environments and jeopardizes the availability factor is denial of service attack (DoS) [6] [5]. Since web services are a combination of a variety of technologies such as SOAP, HTTP, and XML, they are vulnerable to different type of attacks. For example, an attacker sends a malicious request (XML message) to the web service and the XML message forces the XML parser into an infinite recursion exhausting all

Cristian Pinzón
Universidad Tecnológica de Panamá, Av. Manuel Espinosa Batista, Panamá
e-mail: `cristian_ivanp@usal.es`

Angélica González · Manuel Rubio · Javier Bajo
Departamento Informática y Automática
Universidad de Salamanca
Plaza de la Merced s/n, 37008, Salamanca, Spain
e-mail: `{angelica,mprc,jbajope}@usal.es`

E. Corchado et al. (Eds.): SOCO 2010, AISC 73, pp. 1–8.

available computing resources. As a result, the attack prevents access the available services to the authorized users.

Response time is a critical aspect in the majority of internet security systems. This article presents a novel proposal to cope with DoS attacks, but unlike existing solutions [6], [5], [8], [10], [9], [2] our proposal takes into account the different mechanisms that can lead to a DoS attack. In addition, our proposal is based on a real time classifier agent that incorporates a mixture of experts to choose a specific technique of classification depending on the feature of the attack and the available time to solve the classification. The internal structure of the agent is based on the Case-Base Reasoning (CBR) model [3], with the main difference being that the different CBR phases are time bounded, thus enabling its use in real time.

The rest of the paper is structured as follows: section 2 presents the problem that has prompted most of this research work. Section 3 shows a general view of the temporal bounded CBR used as deliberative mechanism in the classifier agent. Section 4 explains in detail the classification model designed. Finally, the conclusions of our work are presented in section 5.

2 DoS Attacks Description

With XML and Web Services the risk of a DoS attack being carried out increases considerably. The most common message protocol for Web Services is SOAP, an XML based message format. Such a SOAP message is usually transported using the HTTP protocol. The DoS attacks at the web services level generally take advantage of the costly process that may be associated with certain types of requests.

Table 1 presents the types of DoS attack analyzed within this study.

Table 1 Types of attacks

Types of Attacks	Description
Recursive Payloads	A message written in XML can harbor as many elements as required, complicating the structure to the point of overloading the parser.
Oversize Payloads	It reduces or eliminates the availability of a web service while the CPU, memory or bandwidth are being tied up by a massive mailing with a large payload.
Buffer overflow	This attack targets the SOAP engine through the Web server. An attacker sends more input than the program can handle, which can cause the service to crash.
XML Injection	Any element that is maliciously added to the XML structure of the message can reach and even block the actual Web service application.
SQL Injection	An attacker inserts and executes malicious SQL statements into XML
XPath Injection	An attacker forms SQL-like queries on an XML document using XPath to extract an XML database.

It is important to understand that the focus of our proposal centers on the classification of web service requests through SOAP messages. Finally, there are several initiatives within this field: [6], [5], [8], [10], [9], [2]. However, the main disadvantage common to each of these approaches is their low capacity to adapt themselves to the changes in the patterns, which reduces the effectiveness of these methods when slight variations in the behaviours of the known attacks occur or when new attacks appear. Moreover, most of the existing approaches are based on a centralized perspective. Because of this and the focus on performance aspects, centralized approaches can become a bottleneck when security is broken, causing a reduction of the overall performance of the application. In addition, none of these approaches considers the limitations or restrictions in the response time.

3 Real Time Agent and Case-Based Reasoning (CBR)

A real time agent is one that is able to support tasks that should be performed within a restricted period of time [7]. Intelligent agents may use a lot of reasoning mechanisms to achieve these capabilities, including planning techniques or Case-Based Reasoning (CBR) techniques. The main assumption in CBR is that similar problems have similar solutions [1]. Therefore, when a CBR system has to solve a new problem, it retrieves precedents from its case-base and adapts their solutions to fit the current situation.

If we want to use CBR techniques as a reasoning mechanism in real-time agents, it is necessary to adapt these techniques to be executed so that they guarantee real-time constraints. In real-time environments, the CBR phases must be temporally bounded to ensure that solutions are produced on time, giving the system a temporally bounded deliberative case-based behaviour. As a first step, we propose a modification of the classic CBR cycle, adapting it so that it can be applied in real-time domains. First, we group the four reasoning phases that implement the cognitive task of the real-time agent into two stages defined as: the learning stage, which consists of the revise and retain phases; and the deliberative stage, which includes the retrieve and reuse phases. Each phase will schedule its own execution time. These new CBR stages must be designed as an anytime algorithm [4], where the process is iterative and each iteration is time-bounded and may improve the final response.

In accordance with this, our Time Bounded CBR cycle (TB-CBR) will operate in the following manner. The TB-CBR cycle starts at the learning stage, checking if there are previous cases waiting to be revised and possibly stored in the case-base. In our model, the solutions provided at the end of the deliberative stage will be stored in a solution list while a feedback about their utility is received. When each new CBR cycle begins, this list is accessed and while there is enough time, the learning stage of those cases whose solution feedback has been recently received is executed. If the list is empty, this process is omitted. After this, the deliberative stage is executed. The retrieval algorithm is used to search the case-base and retrieve a case that is similar to the current case (i.e. the one that characterizes the problem to be solved). Each time a similar case is found, it is sent to the reuse phase where it is transformed into a suitable solution for the current problem by using a reuse algorithm. Therefore, at the end of each iteration of the deliberative stage, the TB-CBR method is able to provide a solution for the problem at hand, although this solution can be improved in subsequent iterations if the deliberative stage has enough time to perform them.

4 Classification Mechanism Based on a Real-Time Agent

This section presents an agent specially designed to incorporate an adaptation of the previously mentioned TB-CBR model as a reasoning engine in which the learning phase is eliminated since it is now performed by human experts. The TB-CBR agent utilizes a global case base, which avoids any duplication of the content of any information compiled from the cases or any information contained in the results of the analysis. Tables 2, 3 and 4 show the structure of the cases. Table 2 shows the fields recovered from the analysis of the service request headers. Table 3 shows the fields associated with the analysis of the service requests that were obtained after the analysis performed by the parser application.

Table 2 Header Fields

Fields	Type	Variable
IDService	Int	h_1
Subnet mask	String	h_2
SizeMessage	Int	h_3
NTimeRouting	Int	h_4
LengthSOAPAction	Int	h_5
TFMessageSent	Int	h_6

Finally, Table 4 shows the information obtained after analyzing the service requests.

From the information contained in the Tables 3 it is possible to obtain the global structure of the cases. In this way, the information of each case can be represented by the following tuple:

$$c = (\{h_i / i = 1...6\} \cup \{p_j / j = 1...27\} \cup \{x_{lm} / l = 1...6, m = 1...2\}) \quad (1)$$

Where l represents the set of existing techniques of attacks shown in Table 1, for each of the values X_l, the first of the parameter stores the probability values obtained by associated technique of attack, while the second parameter stores if it really was an attack.

When the system receives a new request, the TB-CBR agent performs an analysis that can determine whether it is an attack, in which case it identifies the type of attack. The following sections describe the different stages of the deliberative stage for the TB-CBR.

- **Retrieve**

The retrieval time for the cases depends on the size of the cases in the case base. If the size is known, it is easy to predict how much execution time will be used to recover the cases. The asymptotic cost is linear (O(n)). The cases that have

Table 3 Definition of Fields Recovered by the Parser

Description	Fields	Type	Variable
Number of header elements	NumberHeaderElement	Int	p_1
Number of elements in the body	NElementsBody	Int	p_2
Greatest value associated to the nesting elements	NestingDepthElements	Int	p_3
Greatest value associated to the repeated tag within the body	NXMLTagRepeated	Int	p_4
Greatest value associated with the leaf nodes among the declared parents	NLeafNodesBody	Int	p_5
Greatest value of the associated attributes among the declared elements	NAttributesDeclared	Int	p_6
Type of SQL command	Command_Type	Int	p_7
Number of times that the *AND* operator appears in the string	Number_And	Int	p_8
Number of times that the *OR* operator appears in the string	Number_Or	Int	p_9
Number of times the *Group By* function appears	Number_GroupBy	Int	p_{10}
Number of times that the *Order By* function appears	Number_OrderBy	Int	p_{11}
Number of times that the *Having* function appears	Number_Having	Int	p_{12}
Number of *Literals* declared in the string	Number_Literals	Int	p_{13}
Number of times that the *Literal Operator Literal* expression appears	Number_LOL	Int	p_{14}
Length of the SQL string	Length_SQL_String	Int	p_{15}
Greatest value associated with the length of the string among the elements or attributes within the body	LengthStringValueBody	Int	p_{16}
Total number of incidences during the parsing process	TotalNumberIncidenceParsing	Int	p_{17}
Reference to an external entity	URIExternalReference	Int	P_{18}
Number of variables declared in hte XPath expression	XPathVariablesDeclared	Int	P_{19}
Number of elements affected in the consulted node	XpathNumberElementAffected	Int	p_{20}
Number of literals declared in the XQuery Statement	XPathNumberLiteralsDeclared	Int	p_{21}
Number of times the And operator appears in the XQuery Statement	XPathNumberAndOperator	Int	p_{22}
Number of time the *Or* operator appears in the XQuery statement.	XPathNumberOrOperator	Int	p_{23}
Number of functions declared in the XQuery Statement	XPathNumberFunctionDeclared	Int	p_{24}
Lentgh of the XQuery Statement in a SOAP message	XPathLenghtStatement	Int	p_{25}
Cost of processing time (CPU)	CPUTimeParsing	Int	p_{26}
Cost of memory size (KB)	SizeKbMemoryParser	Int	p_{27}

been retrieved during this phase are selected according to the information obtained from the headers of the packages of the HTTP/TCP-IP transport protocol from the new case. The information retrieved corresponds to the service description fields, and the service requestor's subnet mask. Assuming that the newly introduced case is represented by c_{n+1}, the case $c_{n+1}$is defined by the following tuple:

Table 4 Fields Stored as Results

Fields	Type	Variable
Probability Oversize Payload	Real	x_{11}
Attack Oversize Payload	Boolean	x_{12}
Probability Recursive Parsing	Real	x_{21}
Attack Recursive Parsing	Boolean	x_{22}
Probability Buffer Overflow Attack	Real	x_{31}
Attack Buffer Overflow Attack	Boolean	x_{32}
Probability XML Injection Attack	Real	x_{41}
Attack Buffer Overflow Attack	Boolean	x_{42}
Probability Xpath Injection Attack	Real	x_{51}
Attack Xpath Injection Attack	Boolean	x_{52}
Probability SQL Injection Attack	Real	x_{61}
Attack SQL Injection Attack	Boolean	x_{62}

$c_{n+1} = (\{h_i / i = 1...6\})$. The new case does not initially contain information related to the parser since it would be necessary to analyze the content of the message.

$$c_{.h_1h_2} = f_s(C) = \{c_{j.h_1h_2} \in C / c_{j.h_1} = c_{n+1 \cdot h_1} \cap c_{j \cdot h_2} = c_{n+1 \cdot h_2}\} \qquad (2)$$

where $c_{j.h1}$ represents the case j and $h_{1,}$ a property that is determined according to the data shown in Table 2, C represents the set of cases, and f_s the retrieval function. In the event that the retrieved set is empty, the process continues with the retrieval of the messages only without considering the subnet mask.

- **Reuse**

Each type of attack in our proposal (except for Xpath and SQL Injection attacks) can be analyzed by two different techniques. The first is known as the Light technique and is usually a detection algorithm with a low temporal cost, but of low quality as well. On the other hand, using a Heavy technique, the result of the analysis is much more exact, but it requires a much higher amount of execution time. Using these techniques to analyze an attack allows the real time agent to apply the one that is best suited to its needs, without violating the temporal restrictions that should be considered when executing the deliberative stages. In order to determine which is the set of techniques that provides the best solution, it is necessary for the length and quality associated with each technique to be predictable, as seen from a global perspective that includes all possible combinations of techniques.

The different techniques that the classifier agent executes once the optimal combination of techniques has been determined include a set of common inputs that are represented by p_c and are defined as follows: p_c={ p_1, p_{28} , p_{29}, p_2, p_3, p_4 , p_5, p_6, p_{16}, p_{17} , p_{26}, p_{27} }. The remaining entries vary according to the techniques used, which is specified for each one. Table 5 shows the information of the different techniques used for each of the attacks.

Table 5 Techniques Associated with the Attacks

Types of Attacks	Light Techniques	Heavy Techniques
Oversize Payload	Decision Tree	Neural Network
Recursive Parsing	Naïve Bayes	Red neuronal
Buffer Overflow Attack	Decision tree	Red neuronal
XML Injection Attack	SMO	Neural Network
Xpath Injection Attack		Neural Network
SQL Injection Attack		Neural Network

The information used by the different techniques varies according to the type of attack. Table 6 details the different fields associated with each of the attacks for the Heavy techniques. The Light techniques only use the fields that refer to the request headers, specifically the following fields *{h3, h4, h5, h6}*.

Table 6 Inputs Associated with the Different Attack Mechanisms

Types of Attacks	**Variable**
Recursive Parsing	$\{h_3, h_4, h_5, h_6, p_c\}$
Oversize Payload	$\{h_3, h_4, h_5, h_6, p_c\}$
XML Injection Attack	$\{h_3, h_4, h_5, h_6, p_c\}$
Buffer Overflow Attack	$\{h_3, h_4, h_5, h_6, p_c\}$
SQL Injection Attack	$\{p_8, p_9, p_{10}, p_{11}, p_{12}, p_{13}, p_{14}, p_{15}\}$ U $\{h_3, h_4, h_5, h_6, p_c\}$
Xpath Injection Attack	$\{p_{19}, p_{20}, p_{21}, p_{22}, p_{23}, p_{24}\}$ U $\{h_3, h_4, h_5, h_6, p_c\}$

In addition to the techniques displayed in Table 5, there is a global neural network that contains each of the inputs listed in table 6 and that is trained for the entire set of cases. This network will be used in those situations where the response time is critical and it is not possible to check each case individually. At the end of the Reuse stage, the optimal output is selected, corresponding to the maximum values provided by each of the experts, so that if any exceeds a given threshold, the service request is considered to be an attack, and classified as such. Once the analysis is complete, if an attack has been detected, the service request is rejected and is not sent to the respective provider. Subsequently, the result of the analysis is evaluated by a human expert through the revise and retain phase, if it is necessary to store the case associated with the request.

5 Conclusion

This article has described a new approach to protect web services environments against DoS attacks. The proposed approach is based on a time real agent, which has ability to make decisions in real time. The internal structure of the agent is based on a CBR model where the different CBR phases are time bounded, thus enabling its use in real time. Additionally, the adaptation phase in the CBR system that is integrated in the agent proposes a new analysis classification model that is carried out by a mixture of experts. This new model makes it possible to divide the complicated classification task into a series of simple subtasks, so that the fusion of the solutions given by the sub tasks generates the final solution.

Our proposal can be considered as a solid alternative to detect and block DoS attacks. We continue working to achieve a full prototype and then evaluate it within several real environments to probe its effectiveness.

Acknowledgments. This work has been supported by the Spanish Ministry of Science and Innovation TIN 2009-13839-C03-03 and The P.E.P. 2006-2010 IFARHU-SENACYT-Panama.

References

[1] Corchado, J.M., Laza, R., Borrajo, L., Yañez, J.C., Luis, A.D., Valiño, M.: Increasing the Autonomy of Deliberative Agents with a Case-Based Reasoning System. International Journal of Computational Intelligence and Applications 3, 101–118 (2003)

[2] Chonka, A., Zhou, W., Xiang, Y.: Defending Grid Web Services from XDoS Attacks by SOTA. In: IEEE International Conference on Pervasive Computing and Communications, pp. 1–6. IEEE Computer Society, Los Alamitos (2009)

[3] De Paz, J.F., Rodríguez, S., Bajo, J., Corchado, J.M.: Case-based reasoning as a decision support system for cancer diagnosis: A case study. International Journal of Hybrid Intelligent Systems 6, 97–110 (2009)

[4] Dean, T., Boddy, M.S.: An Analysis of Time-Dependent Planning. In: 7th National Conference on Artificial Intelligence, pp. 49–54 (1988)

[5] Gruschka, N., Jensen, M., Luttenberger, N.: A Stateful Web Service Firewall for BPEL. In: IEEE International Conference on Web Services, pp. 142–149 (2007)

[6] Im, E.G., Song, Y.H.: An Adaptive Approach to Handle DoS Attack for Web Services. In: Heidelberg, S.B. (ed.) Intelligence and Security Informatics, pp. 634–635 (2005)

[7] Julian, V., Botti, V.: Developing real-time multi-agent systems. Integrated Computer-Aided Engineering 11, 135–149 (2004)

[8] Padmanabhuni, S., Singh, V., Kumar, K.M.S., Chatterjee, A.: Preventing Service Oriented Denial of Service (PreSODoS): A Proposed Approach. In: IEEE International Conference on Web Services (ICWS 2006), pp. 577–584. IEEE Computer Society, Washington (2006)

[9] Ye, X.: Countering DDoS and XDoS Attacks against Web Services. In: IEEE/IFIP International Conference on Embedded and Ubiquitous Computing, pp. 346–352. IEEE Computer Society, Washington (2008)

[10] Yee, C.G., Shin, W.H., Rao, G.S.V.R.K.: An Adaptive Intrusion Detection and Prevention (ID/IP) Framework for Web Services. In: International Conference on Convergence Information Technology (ICCIT 2007), pp. 528–534. IEEE Computer Society, Washington (2007)

Approaching Real-Time Intrusion Detection through MOVICAB-IDS

Martí Navarro, Álvaro Herrero, Emilio Corchado, and Vicente Julián

Abstract. This paper presents an extension of MOVICAB-IDS, a Hybrid Intelligent Intrusion Detection System characterized by incorporating temporal control to enable real-time processing and response. The original formulation of MOVICAB-IDS combines artificial neural networks and case-based reasoning within a multiagent system to perform Intrusion Detection in dynamic computer networks. The contribution of the *anytime* algorithm, one of the most promising to adapt Artificial Intelligent techniques to real-time requirements; is comprehensively presented in this work.

Keywords: Multiagent Systems, Hybrid Artificial Intelligent Systems, Computer Network Security, Intrusion Detection, Temporal Constraints, Time Bounded Deliberative Process.

1 Introduction

Softcomputing techniques and paradigms have been widely used to build Intrusion Detection Systems (IDSs) [1]. MOVICAB-IDS (MObile VIsualisation Connectionist Agent-Based IDS) has been proposed [2, 3] as a novel IDS comprising a

Martí Navarro · Vicente Julián
Departamento de Sistemas Informáticos y Computación
Universidad Politécnica de Valencia, Camino de Vera s/n, 46022, Valencia, Spain
e-mail: {mnavarro,vinglada}@dsic.upv.es

Álvaro Herrero
Civil Engineering Department, University of Burgos
C/ Francisco de Vitoria s/n, 09006 Burgos, Spain
e-mail: ahcosio@ubu.es

Emilio Corchado
Departamento de Informática y Automática, Universidad de Salamanca,
Plaza de la Merced s/n 37008, Salamanca, Spain
e-mail: escorchado@usal.es

E. Corchado et al. (Eds.): SOCO 2010, AISC 73, pp. 9–18.
springerlink.com

Hybrid Artificial Intelligent System (HAIS) to monitor the network activity. It combines different AI paradigms to visualise network traffic for ID at packet level. This hybrid intelligent IDS is based on a dynamic Multiagent System (MAS) [4], which integrates an unsupervised neural projection model and the Case-Based Reasoning (CBR) paradigm [5] through the use of deliberative agents that are capable of learning and evolving with the environment. A dynamic multi-agent architecture is proposed in this study that incorporates both reactive and deliberative (CBR-BDI agents [6]) types of agents. The proposed IDS applies an unsupervised neural projection model [7] to extract interesting traffic dataset projections and to display them through a mobile visualisation interface.

In other line of things, current approaches involve the application of AI techniques in real-time environments to provide real-time systems with 'intelligent' methods to solve complex problems. There are various proposals to adapt AI techniques to real-time requirements; the most promising algorithms within this field being Anytime [8] and approximate processing [9]. One line of research in Real-Time AI is related to large applications or hybrid system architectures that embody real-time concerns in many components [9], such as Guardian[10], Phoenix [11], or SA-CIRCA [12].

The MOVICAB-IDS approach can be treated as a system where its performance could be notably improved integrating real-time restrictions. Response time [13] is a critical issue for most of the security infrastructure components of an organization. The importance of a smart response on time increases in the case of IDSs. Systems that require a response before a specific deadline, as determined by the system needs, make it essential to monitor execution times. Each task must be performed by the system within a predictable timeframe, within which accurate execution of the given response must be guaranteed. This is the main reason for time-bounding the analytical tasks of MOVICAB-IDS. A key step is the assignation of each pending analysis to available 'Analyzer agents', which is performed by the Coordinator agent. Accordingly, temporal constraints are incorporated in the Coordinator agent that maintains its deliberative capabilities. These problems are discussed in this research in the case of the MOVICAB-IDS Coordinator Agent, which has been modelled as an agent with real-time behaviour in order to improve its performance and achieve a predictable behaviour.

This paper is organized as follows. Section 2 briefly outlines the architecture of MOVICAB-IDS. Section 3 shows how the Coordinator agent in MOVICAB-IDS is upgraded to complete an analysis before a certain deadline. To do so, the Coordinator agent integrates a temporal bounded CBR in its deliberative stage, which is comprehensively described in this section. Section 4 presents experimental results to show the benefits that arise from subjecting different phases of CBR to temporal constraints. Finally, the conclusions and future work are discussed in Section 5.

2 MOVICAB-IDS

As proposed for traffic management [14], different tasks perform traffic monitoring and ID. For the data collecting task, a 4-stage framework [15] is adapted to

MOVICAB-IDS in the following way: (i) **Data capture:** as network-based ID is pursued, the continual data flow of network traffic must be managed. This data flow contains information on all the packets travelling along the network to be monitored; (ii) **Data selection**: NIDSs have to deal with the practical problem of high volumes of quite diverse data [16]. To manage high diversity of data, MOVICAB-IDS splits the traffic into different groups, taking into account the protocol (UDP, TCP, ICMP, and so on) over IP, as there are differences between the headers of these protocols. Once the captured data is classified by the protocol, it can be processed in different ways; (iii) **Segmentation**: The two first stages do not deal with the problem of continuity in network traffic data. The CMLHL model (as some other neural models) can not process data "on the fly". To overcome this shortcoming, a way of temporarily creating limited datasets from this continuous data flow is proposed by segmentation; (iv) **Data pre-processing**: Finally, the different datasets (simple and accumulated segments) must be preprocessed before presenting them to the neural model. At this stage, categorical features are converted into numerical ones. This happens with the protocol information; each packet is assigned a previously defined value according to the protocol to which it belongs.

Once the data-collecting task is performed and the data is ready, the MOVICAB-IDS process performs two further tasks: (v) Data analysis: CMLHL is applied to analyse the data. Some other unsupervised models have also been applied to perform this task for comparison purposes; (vi) Visualisation: the projections of simple and accumulated segments are presented to the network administrator for scrutiny and monitoring. One interesting feature of the proposed IDS is its mobility; this visualisation task may be performed on a different device other than the one used for the previous tasks. To improve the accessibility of the system, results may be visualised on a mobile device (such as phones or blackberries), enabling informed decisions to be taken anywhere and at any time. In summary, the MOVICAB-IDS task organisation comprises the six tasks described above.

MOVICAB-IDS has been designed, on the basis of Gaia methodology [17], [18], as a MAS that incorporates the following six agents:

- **Sniffer:** This reactive agent is in charge of capturing traffic data. The continuous traffic flow is captured and split into segments in order to send it through the network for further processing. Finally, the readiness of the data is communicated. One agent of this class is located in each of the network segments that the IDS has to cover (from 1 to *n*).
- **Preprocessor:** After splitting traffic data, the generated segments are preprocessed prior to their analysis. Once the data has been preprocessed, an analysis for this new piece of data is requested.
- **Analyzer:** This is a CBR-BDI agent. It has a connectionist model embedded in the adaptation stage of its CBR system that helps to analyze the preprocessed traffic data. The connectionist model is called Cooperative Maximum Likelihood Hebbian Learning (CMLHL) [7]. This agent generates a solution (or achieves its goals) by retrieving a case and analyzing the new one using a CMLHL network.

- **ConfigurationManager:** The configuration information is important as data capture, data splitting, preprocessing and analysis depend on the values of several parameters, such as packets to capture, segment length,... This information is managed by the ConfigurationManager reactive agent, which is in charge of providing this information to the Sniffer, Preprocessor, and Analyzer agents.
- **Coordinator:** There can be several Analyzer agents (from 1 to *m*) but only one Coordinator: the latter being in charge of distributing the analyses among the former. In order to improve the efficiency and perform real-time processing, the preprocessed data must be dynamically and optimally assigned. This assignment is performed taking into account both the capabilities of the machines where the Analyzer agents are located and the analysis demands (amount and volume of data to be analysed). As is well known, the CBR life cycle consists of four steps: retrieval, reuse, revision and retention [5].
- **Visualizer:** This is an interface agent. At the very end of the process, the analyzed data is presented to the network administrator (or the person in charge of the network) by means of a functional, mobile visualization interface. To improve the accessibility of the system, the administrator may visualize the results on a mobile device, enabling informed decisions to be taken anywhere and at any time.

3 Time-Bounding the MOVICAB-IDS Coordinator Agent

CBR-BDI agents [19] integrate the BDI (Belief-Desire-Intention) software model and the Case-Based Reasoning (CBR) paradigm. They use CBR systems [5] as their reasoning mechanism, which enables them to learn from initial knowledge, to interact autonomously with the environment, users and other agents within the system, and which gives them a large capacity for adaptation to the needs of its surroundings. These agents may incorporate different identification or projection algorithms depending on their goals. In this case, an ANN will be embedded in such agents to perform ID in computer networks.

The MOVICAB-IDS Coordinator agent, in charge of assigning the pending analyses to the available Analyzer agents, is defined as a Case-Based Planning (CBP-BDI) agent [20]. CBP [21] attempts to solve new planning problems by reusing past successful plans [22]. The Coordinator agent plans to allocate an analysis to one of the available Analyzer agents based on the following criteria:

- **Location.** Analyzer agents located in the network segment where the Visualizer or Pre-processor agents are placed would be prioritised.
- **Available resources** of the computer where each Analyzer agent is running. The computing resources and their rate of use all have to be taken into account. Thus, the work load of the computers must be measured.
- **Analysis demands.** The amount and volume of data to be analysed are key issues to be considered.
- **Analyser agents behaviour.** As previously stated, these agents behave in a "learning" or "exploitation" mode. Learning behaviour causes an Analyzer agent to spend more time over an analysis than exploitation behaviour does.

As a computer network is an unstable environment, the availability of Analyzer agents change dynamically. Network links may stop working from time to time, so the Coordinator agent must be able to re-assign the analyses previously sent to the Analyzer agents located in the network segment that may be down at any one time. As previously stated, the current version of the MOVICAB-IDS is unable to ensure the analysis of a network segment in a maximum amount of time, losing efficiency and reducing the CPU utilization capability. In order to improve the efficiency and perform real-time processing, the Coordinator agent is upgraded to become a Temporal Bounded Case-Based Planning (TB-CBP) BDI agent, bringing MOVICAB-IDS closer to real-time ID. TB-CBP is based on Temporal Bounded CBR as explained in the next section.

3.1 Temporal Bounded CBR

The Temporal Bounded CBR (TB-CBR) is a modification of the classic CBR cycle specially adapted to be applied in domains with temporal constraints. In real-time environments, the CBR stages must be temporal bounded to ensure that the solutions are produced on time; giving the system a temporal bounded deliberative case-based behaviour.

The different phases of the TB-CBR cycle are grouped in two stages according to their function within the reasoning process of an agent with real-time constraints. The fist one, called learning stage, consists of the revise and retain phases; and the second one, named the deliberative stage, includes the retrieve and reuse phases. Each phase will schedule its own execution time to support the designer in the time distribution among the TB-CBR phases. These stages can incorporate an anytime algorithm [23], where the process is iterative and each iteration is time-bounded and may improve the final response.

To ensure up-to-date cases in the case base, the TB-CBR cycle starts at the learning stage, which entails checking whether previous cases are awaiting revision and could be stored in the case base. The solutions provided by the TB-CBR are stored in a solution list at the end of the deliberative stage. This list is accessed when each new TB-CBR cycle begins. If there is sufficient time, the learning stage is implemented for cases where solution feedback has recently been received. If the list is empty, this process is omitted.

Once the learning stage finishes, the deliberative starts. The retrieval algorithm is used to search the case base and chose a case that is similar to the current case (i.e. the one that characterizes the problem to be solved). Each time a similar case is found, it is sent to the reuse phase where it is transformed into a suitable plan for the current problem by using a reuse algorithm. Therefore, at the end of each iteration in the deliberative stage, the TB-CBR method is able to provide a solution to the problem at hand, which may be improved in subsequent iterations if there is any time remaining at the deliberative stage. See more details in [24].

3.2 TB-CBR Operation within the Coordinator Agent

In the aforementioned environment, analysis planning must be completed within a maximum time. For this reason, an agent, which provide the necessary control mechanisms to carry out this task, is deployed to complete the analysis on time. Consequently, when a new segment is ready for analysis, the Coordinator agent, which is a real-time agent, has a limited amount of time to assign the pending analysis to the available Analyzer agents, which have to provide an answer as soon as possible. Therefore, a temporal constraint on the process (starting with a new generated segment and ending with the Analyzer agent giving the answer) is essential to ensure prompt execution. To perform this temporal control, all the steps in the process must be known and must be temporal bounded. Additionally, the system has to be deterministic. To guarantee these conditions, the Coordinator agent takes advantage of the TB-CBP to assign the pending analysis. So, the four phases of the TB-CBP cycle of the Coordinator agent are re-defined to comply with the temporal constraints following the TB-CBR guidelines (see [24]).

The first stage (learning stage) is executed if the agent has the plans from previous executions stored in the *solutionQueue* (these are previous executions of the CBR cycle that have not been revised and retained). The plans are stored just after the end of the deliberative stage. In this case, the following phases are executed:

- **Revise:** The plan revision consists of a two-fold analysis. On the one hand, planning failures are identified by finding under-exploited resources. As an example, the following hypothetical situation is identified as a planning failure: one of the Analyzer agents is not busy performing an analysis while the other ones have a list of pending analyses. On the other hand, execution failures are detected when communication with Analyzer agents has been interrupted. Information on these failures is stored in the case base for future consideration. When an execution failure is detected, the CBP cycle is run from the beginning, which renews the analysis request.
- **Retain:** When a plan is adopted, the Coordinator agent stores a new case containing the dataset-descriptor and the solution.

The deliberative stage is only launched if there is a new network segment to be analysed (a new pre-processed dataset is ready) by adding it to the *problemQueue* of the Coordinator agent. This will launch the execution of the following phases:

- **(Plan) Retrieve:** As previously stated, when a new pre-processed dataset is ready, an analysis is requested from the Coordinator agent. The most similar plan is obtained by associative retrieval, taking into account the case/plan description. As the time required to extract a case from the case base is predictable, the Coordinator agent knows how long it takes to get the first solution. Moreover, if the Coordinator agent has some extra time to plan the analyses, it will attempt to improve this first plan within the available time by continuing searching previously stored plans.
- **Reuse:** The retrieved plan is adapted to the new planning problem. The only restriction is that the analyses running at that time (the results of which have not yet been reported) cannot be reassigned. The others (pending) can be

reassigned in order to optimize overall performance. This phase is also temporal bounded. The Coordinator agent knows when it will finish the adaptation of the cases to the new planning problem. In this phase, as the Coordinator agent calculates when the analysis agents will finish their tasks, it either knows the available time to continue building the plan. The Analyzer agents will still be executing pending analyses when this phase is completed. Thus, the assignment of an analysis to an Analyzer depends on its work load at that particular time.

The main advantage of using the TB-CBP with regard to using a CBP without temporal constraints is to ensure a system response on time. The use of TB-CBP allows the distribution of the analysis to the Analyzer agents taking into account the available time to perform this task. On the other hand, the application of TB-CBP improves the CPU utilization and minimizes the average execution time of the analyses as it has been checked in a set of tests. The analysis requests are launched for 2 minutes following exponential distribution in which α parameter value is 0.3 (a request is generated each 3 seconds approximately). The results obtained after one hundred executions are shown in Table 1.

Table 1 TB-CBP vs. CBP

	CPU utilization	Analysis fulfilled on time	Average Execution Time
TB-CBP	97 %	98.2 %	1.6 ms
CBP	72 %	61.5 %	2.4 ms

4 Experimental Results: MOVICAB-IDS Visualizations

There are two main dangerous anomalous situations related to SNMP [25].: MIB information transfers and port sweeps or scans. The MIB (Management Information Base) can be defined in broad terms as the database used by SNMP to store information about the elements that it controls. A transfer of some or all the information contained in the SNMP MIB is potentially quite a dangerous situation. A port scan may be defined as series of messages sent to different port numbers to gain information on its activity status.

The effectiveness of MOVICAB-IDS in facing some anomalous situations has been widely demonstrated in previous works [2, 3, 26, 27]. It identifies anomalous situations due to the fact that these situations do not tend to resemble parallel and smooth directions (normal situations) or because their high temporal concentration of packets. It can be seen in Fig. 1.a, where 3 port sweeps have been identified (Group 1) and visualized in a mobile platform. On the other hand, a more advanced visualization is offered in Fig. 1.b for a different data set. In this case, it is easy to notice some different directions (Groups A and B) to the normal data ones. Also, the density of packets is higher for these anomalous groups related to a MIB information transfer.

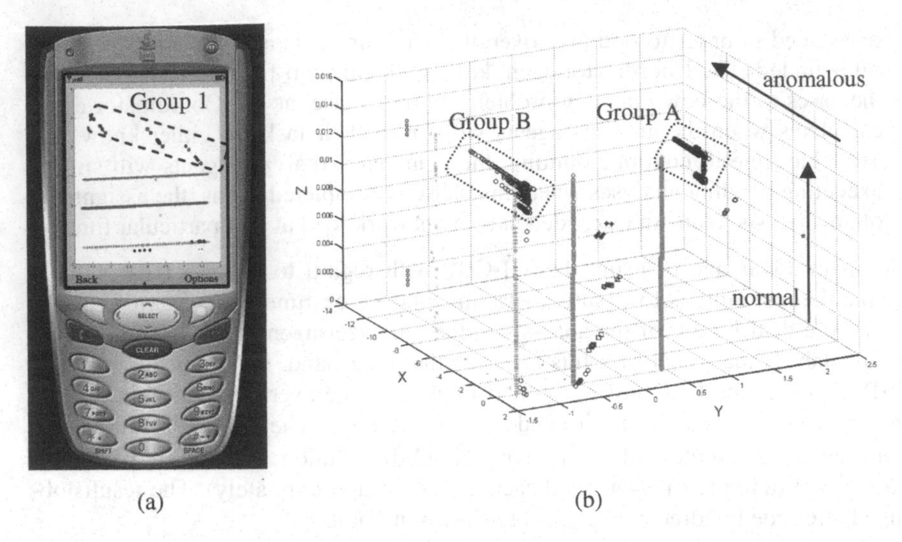

(a) (b)

Fig. 1 Mobile (a) and advanced (b) visualizations provided by MOVICAB-IDS

5 Conclusions and Future Work

An upgraded version of MOVICAB-IDS is presented in this paper. This version imposes temporal constraints on the deliberative agents within a CBR architecture, which enables them to respond to events in real (both hard or soft) time. In this case, the deliberative Coordinator agent, working at a high level with Belief-Desire-Intention (BDI) concepts, is temporal bounded by redefining the four phases of its CBP cycle. The consequences of temporal bounding these phases are described in this paper. As a result, the Coordinator agent will always give a solution within the available time, thereby maximizing CPU utilization and minimizing average execution time of the analyses.

Acknowledgments. This research is funded through the Junta of Castilla and León (BU006A08); the Spanish Ministry of Education and Innovation (CIT-020000-2008-2 and CIT-020000-2009-12); the Spanish government (TIN2005-03395 and TIN2006-14630-C03-01), FEDER and CONSOLIDER-INGENIO (2010 CSD2007-00022). The authors would also like to thank the vehicle interior manufacturer, Grupo Antolin Ingenieria S.A. for supporting the project through the MAGNO2008 - 1028.- CENIT Project funded by the Spanish Ministry of Science and Innovation.

References

1. Abraham, A., Jain, R., Thomas, J., Han, S.Y.: D-SCIDS: Distributed Soft Computing Intrusion Detection System. Journal of Network and Computer Applications 30(1), 81–98 (2007)

2. Herrero, Á., Corchado, E.: Mining Network Traffic Data for Attacks through MOVICAB-IDS. In: Foundations of Computational Intelligence. Studies in Computational Intelligence, vol. 4, pp. 377–394. Springer, Heidelberg (2009)
3. Corchado, E., Herrero, Á.: Neural Visualization of Network Traffic Data for Intrusion Detection. Applied Soft Computing (Accepted with changes) (2010)
4. Wooldridge, M., Jennings, N. R.: Agent theories, architectures, and languages: A survey. Intelligent Agents (1995)
5. Aamodt, A., Plaza, E.: Case-Based Reasoning - Foundational Issues, Methodological Variations, and System Approaches. AI Communications 7(1), 39–59 (1994)
6. Carrascosa, C., Bajo, J., Julián, V., Corchado, J.M., Botti, V.: Hybrid Multi-agent Architecture as a Real-Time Problem-Solving Model. Expert Systems with Applications: An International Journal 34(1), 2–17 (2008)
7. Corchado, E., Fyfe, C.: Connectionist Techniques for the Identification and Suppression of Interfering Underlying Factors. International Journal of Pattern Recognition and Artificial Intelligence 17(8), 1447–1466 (2003)
8. Dean, T., Boddy, M.: An Analysis of Time-dependent Planning. In: 7th National Conference on Artificial Intelligence (1988)
9. Garvey, A., Lesser, V.: A Survey of Research in Deliberative Real-time Artificial Intelligence. Real-Time Systems 6(3), 317–347 (1994)
10. Hayes-Roth, B., Washington, R., Ash, D., Collinot, A., Vina, A., Seiver, A.: Guardian: A Prototype Intensive-care Monitoring Agent. Artificial Intelligence in Medicine 4, 165–185 (1992)
11. Howe, A.E., Hart, D.M., Cohen, P.R.: Addressing Real-time Constraints in the Design of Autonomous Agents. Real-Time Systems 2(1), 81–97 (1990)
12. Musliner, D.J., Durfee, E.H., Shin, K.G.: CIRCA: A Cooperative Intelligent Real-time Control Architecture. IEEE Transactions on Systems, Man, and Cybernetics 23(6), 1561–1574 (1993)
13. Kopetz, H.: Real-time Systems: Design Principles for Distributed Embedded Applications. Kluwer Academic Publishers, Dordrecht (1997)
14. Babu, S., Subramanian, L., Widom, J.: A Data Stream Management System for Network Traffic Management. In: Workshop on Network-Related Data Management, NRDM 2001 (2001)
15. Herrero, Á., Corchado, E.: Traffic Data Preparation for a Hybrid Network IDS. In: Corchado, E., Abraham, A., Pedrycz, W. (eds.) HAIS 2008. LNCS (LNAI), vol. 5271, pp. 247–256. Springer, Heidelberg (2008)
16. Dreger, H., Feldmann, A., Paxson, V., Sommer, R.: Operational Experiences with High-Volume Network Intrusion Detection. In: 11th ACM Conference on Computer and Communications Security. ACM Press, New York (2004)
17. Zambonelli, F., Jennings, N.R., Wooldridge, M.: Developing Multiagent Systems: the Gaia Methodology. ACM Transactions on Software Engineering and Methodology 12(3), 317–370 (2003)
18. Wooldridge, M., Jennings, N.R., Kinny, D.: The Gaia Methodology for Agent-Oriented Analysis and Design. Autonomous Agents and Multi-Agent Systems 3(3), 285–312 (2000)
19. Pellicer, M.A., Corchado, J.M.: Development of CBR-BDI Agents. International Journal of Computer Science and Applications 2(1), 25–32 (2005)
20. Bajo, J., Corchado, J., Rodríguez, S.: Intelligent Guidance and Suggestions Using Case-Based Planning. In: Weber, R.O., Richter, M.M. (eds.) ICCBR 2007. LNCS (LNAI), vol. 4626, pp. 389–403. Springer, Heidelberg (2007)

21. Hammond, K.J.: Case-based Planning: Viewing Planning as a Memory Task. Academic Press Professional, Inc., London (1989)
22. Spalzzi, L.: A Survey on Case-Based Planning. Artificial Intelligence Review 16(1), 3–36 (2001)
23. Dean, T., Boddy, M.S.: An Analysis of Time-Dependent Planning. In: 7th National Conference on Artificial Intelligence (1988)
24. Navarro, M., Heras, S., Julián, V.: Guidelines to Apply CBR in Real-Time Multi-Agent Systems. Journal of Physical Agents 3(3), 39–43 (2009)
25. Case, J., Fedor, M.S., Schoffstall, M.L., Davin, C.: Simple Network Management Protocol (SNMP). IETF RFC 1157 (1990)
26. Corchado, E., Herrero, Á., Sáiz, J.M.: Detecting Compounded Anomalous SNMP Situations Using Cooperative Unsupervised Pattern Recognition. In: Duch, W., Kacprzyk, J., Oja, E., Zadrozny, S. (eds.) ICANN 2005. LNCS, vol. 3697, pp. 905–910. Springer, Heidelberg (2005)
27. Corchado, E., Herrero, Á., Sáiz, J.M.: Testing CAB-IDS Through Mutations: On the Identification of Network Scans. In: Gabrys, B., Howlett, R.J., Jain, L.C. (eds.) KES 2006. LNCS (LNAI), vol. 4252, pp. 433–441. Springer, Heidelberg (2006)

Hybrid Dynamic Planning Mechanism for Virtual Organizations

Sara Rodríguez, Vivian F. López, and Javier Bajo

Abstract. It is possible to establish different types of agent organizations according to the type of communication, the coordination among agents, and the type of agents that comprise the group. Each organization needs to be supported by a coordinated effort that explicitly determines how the agents should be organized and carry out the actions and tasks assigned to them. This paper presents a new global coordination model for an agent organization. This model is unique in its conception, allowing an organization in a highly dynamic environment to employ self-adaptive capabilities in execution time.

1 Introduction

Ideally, MAS include the following characteristics [10].:(i) They are typically open with a non-centralized design. (ii) They contain agents that are autonomous, heterogeneous and distributed, each with its own "personality" (cooperative, selfish, honest, etc.). (iii) They provide an infrastructure specifically for communication and interaction protocols. Open MAS should allow the participation of heterogeneous agents with different architectures and even different languages [14]. However, this makes it impossible to trust agent behavior unless certain controls based on norms or social rules are imposed. To this end, developers have focused on the organizational aspects of agent societies, using the concepts of organization, norms, roles, etc. to guide the development process of the system.

Virtual organizations [6] are a means of understanding system models from a sociological perspective. From a business perspective, a virtual organization model is based on the principles of cooperation among businesses within a shared network, and exploits the distinguishing elements that provide the flexibility and quick response capability that form the strategy aimed at customer satisfaction.

Sara Rodríguez · Vivian F. López · Javier Bajo
Departamento Informática y Automática
Universidad de Salamanca
Plaza de la Merced s/n, 37008, Salamanca, Spain
e-mail: {srg,vivian,jbajope}@usal.es

E. Corchado et al. (Eds.): SOCO 2010, AISC 73, pp. 19–26.

Even so, within the development of organizations, both at the business and agent level, we find a set of requirements [12] that call for the use of new social models in which the use of open and adaptive systems is possible [14].

Given the advantages provided by the unique characteristics found in the development of MAS from an organizational perspective, and the absence of an adaptive planning process for any social model, this study proposes a model that can coordinate a dynamic and adaptive planning system in an agent organization. The article is structured as follows: Section 2 describes the state of the art for current studies of the agent organizations and its adaptation. Section 3 presents the proposed planning model. Section 4 demonstrates how the model can be used in a case study and shows some results and conclusions obtained.

2 Background

There are several different organizational approaches [4][14]. However, while these studies provide mechanisms for creating coordination among participants, there is much less work focused on adapting organizational structures in execution time or norms defined in design time. For example, [9] proposes a model for controlling adaption by creating new norms. [7] propose a distributed model for reorganizing their architecture. [1] requires agents to follow a protocol to adapt the norms. Each of these studies focuses on the structure and/or norms based on adapting the coordination among participants. Another possibility is the development of a MAS that focuses on the concept of organization/institution. One electronic institution [5] should be considered a social middleware between the external participating agents and the selected communication layer responsible for accepting or rejecting the agent actions. The primary difference with the other proposals is that the adaption is carried out by the institution instead of by the agents. Lastly, there are approaches focus on social group mechanisms based on the social information gathered during the interactions [13].

None of these approaches is capable of coordinating tasks for the member agents of the organization to solve a common problem, nor do them consider that task planning should adapt to changes in the environment. The social model used in the architecture selected for this study is THOMAS [2][8], which focuses on defining the structure and norms. The following section will present the planning model proposed integrated into THOMAS whose goal is to carry out an adaptive planning process within an agent organization.

3 Planning Model

In this research is proposed a planning model that facilitates a self-adaptation feature within an agent society. We will use a cooperative MAS in which each agent is capable of establishing plans dynamically in order to reach its objectives. The global mechanism considers the global objective of the society, as well as its norms and roles. It's obtained a planning model that can, within an architecture geared towards the development of agent organizations (THOMAS [2][8]), take into account the changes that are produced within an environment during the execution of a plan. The planning process defines the actions that the society of agents will have to execute and should therefore also take into account the particular circumstances of each of its members. To achieve this, a CBP-BDI (Case Based Planning) agent is used, applying the planning model showed in this section, that is particularly suited for organizations. A CBP-BDI agent is a specialization of an CBR-BDI agent [3]. A CBP-BDI agent calculates the plan or intention that is most easy to replan: Most RePlannable Intention (MRPI). This is the plan that can most easily be replaced by another plan in case it is interrupted (for example, if a user changes preferences while the plan is being executed.

A plan p within an organization is defined as p=<E, O, O', R, R'>, where: E is the environment that represents the type of problem that the organization solves, and is characterized by a set of states $E = \{e_0, e^*\}$ for each agent, where e_0 represents the initial state of the agent when the plan begins, y and e^* is the state or set of states that the agents tries to achieve. O represents the set of objectives for the individual agent and O' is the set of objectives reached once the plan has been executed. R is the set of available resources for the given agent and R' is the set of resources that the agent has used during the execution of the plan.

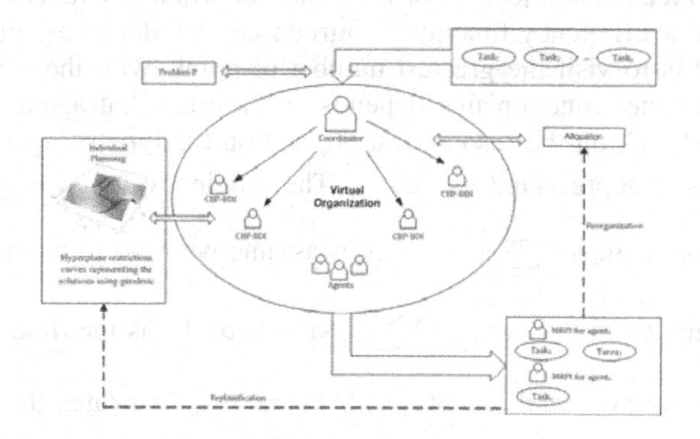

Fig. 1 Planning Model

Given the initial state of the organization, the term *global planning* is used to describe the search for a solution that can reach the final state, all the while complying with a series of requirements for the organization. The problem can be represented in a planning space that is delimited by the restrictions imposed by the requirements. Given a common objective, specified resources available and tasks to perform, the aim is to find a global plan that allows the organization to find the optimal solution, To this end, the planning agent should bear in mind the optimal plans $p^*(t)$ obtained for each individual agent. It is not necessary for all of the agents within the organization to know how to meet the objectives, but they should know how to perform some of the tasks that contribute towards reaching those objectives for the organization.

Upon initiating the process, certain agents will be retrieved from the data memory of cases to perform at least one of the problem tasks. For each task that is not completed by any of the retrieved agents, at least one new agent will be incorporated. This agent will have the greatest probability of successfully completing the given task. The idea is to count on the necessary agents so that no task is left unassigned.

Let us assume that the common objective for agents "m" has "n" states or tasks with ss $m, n \in \mathbb{N}$. Each agent has its own characteristics with regards to which tasks it can perform, which resources to use, and the amount of time available to perform the tasks. In other words, each agent has its own profile. Given a state "j" for each agent "i" where $i \in \{1, \cdots, m\}_{m \in \mathbb{N}}$, it can be defined with a tuple z_{ij} - where each coordinate in the tuple refers to the characteristic that defines it.

The following binary variables are defined as:

$$a_{ij} = \begin{cases} 1 \, \textit{if agent "i" is assigned to task "j"} \\ 0 \, \textit{otherwise} \end{cases}$$

For each problem related to assigning tasks, an objective function is defined whose goal is to minimize and maximize the cost used by agents "m" to perform the common objective. For example, minimize or maximize the cost of using one of the agents to reach an objective, or maximize an efficiency function as need for each case. A new efficiency function is introduced in order to assign tasks to the agents. Its aim is to visit the greatest number of points with the lowest possible cost. Cost is another function that depends on the time that agent "i" has spent working on task "j", on the resources used, and on the type of agent assigned to each task. This is represented as: $c^{i}_{t_{ij} r_{ij}}$ .The efficiency function is defined as: $\textit{Efficiency} = N^{o} \textit{ points visited} / \sum_{i=1}^{m} \sum_{j=1}^{n} c^{i}_{t_{ij} r_{ij}} \cdot a_{ij}$ .Let us assume we want to maximize the efficiency function: $\textit{Max} \cdot N^{o} \textit{ points visited} / \sum_{i=1}^{m} \sum_{j=1}^{n} c^{i}_{t_{ij} r_{ij}} \cdot a_{ij}$ where t_{ij} is the time it takes agent "i" to perform the task, and $t_{ij} = \underset{k}{Máx}\{t_{ijk}\}$ where t_{ijk} indicates the time it takes agent "i" to perform task "j" for tourist "k". Taking the maximum value of "k" (type of tourist), we can ensure that the guide has time to perform the necessary task regardless of the type of tourist. These times are initially estimated. Let us now define the restrictions of the problem.

1. We want each state to be completed by an agent, which in mathematical terms can be stated, for each state "k" as: $\sum_{i=1}^{m} a_{ik} = 1 \quad \forall k \in \{1, \cdots, n\}$

2. We want each state to be completed within a specified period of time. Let us assume that state "k" should be completed within time t_k. The restriction would be:. $\sum_{i=1}^{m} t_{ik} a_{ik} \leq t_k \quad \forall k \in \{1, \cdots, n\}$

3. Each state "k" needs a set of resources to be executed. There is no reason for all of the agents to have these resources.

Given state "k", we need r_h^k resources with $h \in \mathbb{N}$, where $r_w = max\{r_h^k\}_{h \in \mathbb{N}}^{k=1,\cdots,n}$.

The variables $\{r_x^k\}_{x \in \{1,\cdots,w\}}$ $\forall k \in \{1,\cdots,n\}$ are defined in binary form:

$$r_x^k = \begin{cases} 1 \text{ if the agent "k" needs the resource "x"} \\ 0 \text{ otherwise} \end{cases}$$

The agent that performs state "k" must at the very least have at its disposable the resources that are needed to perform state "k", for which, given state "k", for each resource from the set $\{r_x^k\}_{x \in \{1,\cdots,w\}}$ $\forall k \in \{1,\cdots,n\}$ we can define the following restriction:. $\sum_{i=1}^{m} r_{ix} a_{ik} \geq r_x^k \quad \forall k \in \{1,\cdots,n\}; \forall x \in \{1,\cdots,w\}$. The variables $\{r_{ix}\}_{x \in \{1,\cdots,w\}}$ $\forall i \in \{1,\cdots,m\}$ are binary variables: $r_{ix} = \begin{cases} 1 \text{ if the agent "i" has the resource "x"} \\ 0 \text{ otherwise} \end{cases}$

4. Each agent "i" has a minimum and maximum time for work, depending on the type of agent. These times are represented as $t_i^{Turn\ on}$ and $t_i^{Turn\ off}$ respectively: $t_i^{Turn\ on} \leq \sum_{j=1}^{n} t_{ij} \leq t_i^{Turn\ off} \quad \forall i \in \{1,\cdots,m\}$ For the majority of agents, as we will see in the case study, the maximum number of working hours is equal to a regular 8 hour work day.

5. Every time we assign tasks to an agent, we want it to perform the minimum number of tasks, which varies according to the type of agent: $\sum_{j=1}^{n} a_{ij} \geq NumberTask_i \quad \forall i \in \{1,\cdots,m\}$. If the suggested problem of non-linear programming were incompatible, we would add agents to make it compatible. The agent added would be the one with the highest probability a priori of performing the necessary tasks. If a norm (restriction) changes, it would be necessary to assign tasks once again. This allows us to obtain a plan for the tasks that need to be performed by the agent organization. In other words, we can obtain a global plan composed of all the tasks and agents in the organization that will carry them out. Every agent in the organization recognizes the tasks that it needs to perform. These agents, which are CBP-BDI agents, integrate the 4 phases of a CBR system (retrieval, reuse, revise and retain).

4 Case Study and Experimental Results

This section presents a case study that tests the defined model. An organization is implemented by using the model proposed in section 3 and is represented in a virtual world [11] containing a set of cultural heritage sites. The simulation within the virtual world represents a tourist environment in which there are guides and tourists, and in which the tour guide´s tasks will be performed in adherence to a defined set of norms. The roles that have been identified within the case study are: *Tourist*, *Monument*, *Guide*, *Visitor*, *Coordinator*, *Notification* and *Manager*: The agents that take on the role of Guide are those that will carry out dynamic planning according to the tasks they need to carry out for each group of tourists. The generated plans should ensure that all of the visitors assigned to a tour guide are able to follow their tourist route. They will be personalized according to the Guide´s profile and work habits, and should take into account the restrictions directly related to each agent on an individual basis, as well as the restrictions of the organization itself. These restrictions are imposed according to the norms for the society of agents: (i) the work schedule for a Guide agent (8 hours); (ii) the maximum number of Tourist agents assigned to a guide; (iii) visiting days and hours for certain monuments; (iv)the maximum number of Guide agents that can participate on a route; (v) the minimum number of points to visit on a route.

Once the Coordinator has identified all of the agents in the organization that are needed to carry out the plan, it assigns each task to the agent responsible for completing it. At that moment each Guide agent becomes aware of its tasks and designs an individual plan. Each Guide agent is a type of CBP-BDI agent capable of providing efficient plans in execution time. The following paragraph provides a detailed example.

Let $E_g = \{e_0^g, \cdots, e_n^g\}$ be the tasks carried out by a group of tourists and visitors "g" in order of priority. We have the following problem $E = \bigcup_g E_g = \{e_0, \cdots, e_n\}$, where E represents the complete set of tasks that must be completed (for this reason they are not superscripted). Let us assume there are 10 guides. Randomly selecting a Guide i∈{1,• , 10}, (specifically, i=3), the task assignment according to their profile is: (1) Agent Task: Visit the cathedral with tourist group 2 $\equiv e_1^2$; t_{31}=30 min. (2) Agent Task: Take tourist group 2 to the aqueduct $\equiv e_2^2$; t_{32}=15 min. (3) Agent Task: Take tourist group 2 to the hermitage $\equiv e_3^2$; t_{33}=10 min. (4) Agent Task: Visit the hermitage $\equiv e_4^2$; t_{34}=10 min. (5) Agent Task: Take tourist group 2 to the Roman city $\equiv e_5^2$; t_{35}=20 min. (6) Agent Task: Visit the Roman city $\equiv e_6^2$; t_{36}=30 min. (7) Agent Task: Take tourist group 2 to the ravine $\equiv e_7^2$; t_{37}=50 min. (8) Agent Task: Hike along the ravine with group 2 $\equiv e_8^2$; t_{38}=20 min. (9) Agent Task: Return to the cathedral with group 2$\equiv e_9^2$; t_{39}=10 min. Calculating the assigned tasks ensures both that the total amount of time assigned to a Guide does not exceed 8 hours, and that any other restrictions corresponding to the norms of the organization are also respected. Each task has a set of objectives that must be met so that the global plan can be successfully completed. To perform each task, the Guide agent should have the number of available resources. For example, the task "Buy tickets for museum 1" corresponds to the objective "Visit museum 1"$\equiv O_0$ and "breakfast, lunch, tea and dinner" correspond to the objective $\equiv O_{2,4,6,7}$ (task 2 indicates breakfast, task 4 indicates lunch, task 6 indicates tea, and 7 indicates dinner). A similar coding is used for resources. As shown in Fig. 2a, value 1 indicates the resource that is needed or the objective to be met, while zero denotes the contrary. Fig. 2a shows the representation of a space $\Re^3$ for tasks according to the following three coordinates: time, number of objectives achieved, and number of resources used (coordinates taken from similar retrieved cases). Specifically, Fig. 2a shows a hyper plan of restrictions and the plan followed for a case retrieved from the beliefs base, considered to be similar to the new case. There are 120 possible routes, not all of which are viable because of the previously mentioned restrictions. In a simulated scenario where the Coordinator assigned this group of tourists to a Guide, the planning process used by the Guide for the tasks it needed to perform is the same as that shown in Fig 2a.

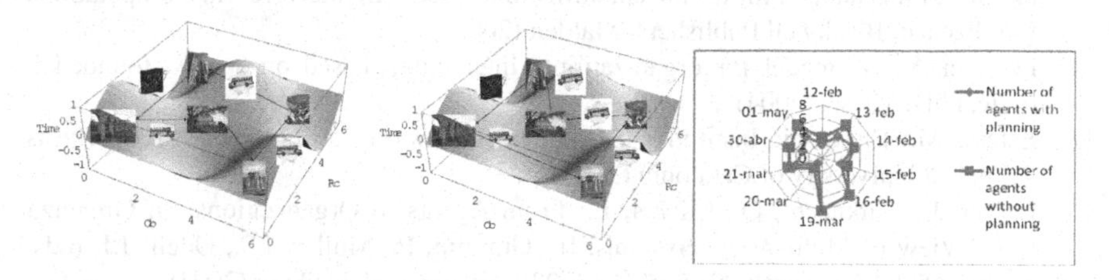

Fig. 2 a) Representation of a space $\Re^3$ for tasks (a) and replanned tasks (b). Number of agents working simultaneously (c).

Figure 2 illustrates the plan as it was carried out. To understand the graphical representation, let us focus on the initial task e_1 and the final task e_9. In between these two tasks, the Guide agent could carry out other tasks that would involve the same or different tourists and visitors. The idea presented in the planning model is to select the optimal plan, the one with the most plans surrounding it, as the solution. The following studies were carried out: Given the same tourists attractions to be visited on the same day, and the same number of tourists per group, one group used the planner and the other did not.

The results for different days, as far as the number of Guide agents used, can be observed in Fig.2c. The color blue represents the average number of guides needed each day using the planner, and red the number without using it. The proposed model helps the organization utilize fewer guides, thus minimizing its costs.

In conclusion, we can affirm to have achieved out stated objectives: (i) Develop agent societies; (ii) Simulate the behavior of an organization in a specific case involving the coordination and adaption of its agents; and (iii) Validate the proposed planning model through a simulation of the organization in a case study. As previously mentioned, it is increasingly common to model a MAS not only from the perspective of the agent and its communication capabilities, but by including organizational engineering as well.

Acknowledgments. This work has been supported by the JCYL project SA071A08.

References

[1] Artikis, A., Kaponis, D., Pitt, J.: Multi-Agent Systems: Semantics and Dynamics of Organisational Models. In: Dynamic Specifications of Norm-Governed Systems, IGI Globa (2009)

[2] Carrascosa, Giret, C.A., Julian, V., Rebollo, M., Argente, E., Botti, V.: Service Oriented MAS: An open architecture (Short Paper). In: Decker, Sichman, Sierra, Castelfranchi (eds.) Proc. of 8th Int. Conf. on Autonomous Agents and Multiagent Systems (AAMAS 2009), Budapest, Hungary, May 10–15, vol. Sierra, pp. 1291–1292 (2009)

[3] Corchado, J.M., Glez-Bedia, M., de Paz, Y., Bajo, J., y de Paz, J.F.: Concept, formulation and mechanism for agent replanification: MRP Architecture. In: Computational Intelligence. Blackwell Publishers, Malden (2008)

[4] Dignum. V.: A model for organizational interaction: based on agents, founded in logic, PhD. Thesis (2004)

[5] Esteva, M.: Electronic Institutions: from specification to development Ph.D. Thesis, Technical University of Catalonia (2003)

[6] Ferber, J., Gutknecht, O., Michel, F.: From Agents to Organizations: an Organizational View of Multi-Agent Systems. In: Giorgini, P., Müller, J.P., Odell, J.J. (eds.) AOSE 2003. LNCS, vol. 2935, pp. 214–230. Springer, Heidelberg (2004)

[7] Gasser, L., Ishida, T.: A dynamic organizational architecture for adaptive problem solving. In: Proc. of AAAI 1991, pp. 185–190 (1991)

[8] Giret, V., Julian, M., Rebollo, E., Argente, C., Carrascosa, Botti, V.: An Open Architecture for Service-Oriented Virtual Organizations. In: Seventh international Workshop on Programming Multi-Agent Systems. PROMAS 2009, pp. 23–33 (2009)

[9] Hubner, J.F., Sichman, J.S., Boissier, O.: Using the Moise+ for a cooperative framework of mas reorganisation. In: Bazzan, A.L.C., Labidi, S. (eds.) SBIA 2004. LNCS (LNAI), vol. 3171, pp. 506–515. Springer, Heidelberg (2004)

[10] Huhns, M., Stephens, L.: Multiagent Systems and Societies of Agents. In: Weiss, G. (ed.) Multi-agent Systems: a Modern Approach to Distributed Artificial Intelligence, MIT, Cambridge (1999)

[11] `http://repast.sourceforge.net` (2009)

[12] Rodríguez, S., Pérez-Lancho, B., De Paz, J.F., Bajo, J., Corchado, J.M.: Ovamah: Multiagent-based Adaptive Virtual Organizations. In: 12th International Conference on Information Fusion, Seattle, Washington, USA, Julio (2009)

[13] Villatoro, D., Sabater-Mir, J.: Categorizing Social Norms in a Simulated Resource Gathering Society. In: Proceedings of the AAAI Workshop on Coordination, Organizations, Institutions and Norms, COIN @ AAAI 2008 (2008)

[14] Zambonelli, F., Jennings, N.R., Wooldridge, M.: Developing Multiagent Systems: The Gaia Methodology. ACM Transactions on Software Engineering and Methodology 12, 317–370 (2003)

Combinatorial Auctions for Coordination and Control of Manufacturing MAS: Updating Prices Methods

Juan José Lavios Villahoz, Ricardo del Olmo Martínez, and Alberto Arauzo Arauzo

Abstract. We use the paradigm of multiagent systems to solve the Job Shop problem. It concerns the allocation of machines to operations of some production process over time periods and its goal is the optimization of one or several objectives. We propose a combinatorial auction mechanism to coordinate agents. The "items" to be sold are the time slots that we divide the time horizon into. In tasks scheduling problems tasks need a combination of time slots of multiple resources to do the operations. The use of auctions in which different valuations of interdependent items are considered (e.g. combinatorial auctions) is necessary. The auctioneer fixes prices comparing the demand over a time slot of a resource with the capacity of the resource in this time slot. Our objective is to find an updating price method for combinatorial auctions that meet the needings of scheduling manufacturing systems in dynamic environments, e.g. robustness, stability, adaptability, and efficient use of available resources.

1 Multiagent Systems in Manufacturing

Manufacturing and production systems are one of the most known fields of application of Scheduling problems. It concerns the allocation of resources to tasks over time periods and its goal is the optimization of one or several objectives. Tasks are operations of some production process and resources are machines in a workshop. [1]. The Manufacturing scheduling problem is featured by its highly combinatorial and dynamic nature and its practical interest for industrial applications [2]. Multiagent Systems have proved to be an appropriate paradigm to model manufacturing and production systems because of their autonomous, distributed and dynamic nature,

Juan José Lavios Villahoz · Ricardo del Olmo Martínez
INSISOC. Escuela Politécnica Superior, Universidad de Burgos, Spain
e-mail: {jjlavios,rdelolmo}@ubu.es

Alberto Arauzo Arauzo
INSISOC. ETSII, Universidad de Valladolid, Spain
e-mail: arauzo@insisoc.org

E. Corchado et al. (Eds.): SOCO 2010, AISC 73, pp. 27–30.
springerlink.com

and robustness against failures. They constitute a useful frame to define distributed decision-making processes. A revision of the main work in this area can be found in [3, 4, 5]. In general, agents are used to encapsulate physical and logical entities or even functionalities of the production system, building systems that are based on the autonomy of each agent and on the interaction and negotiation among them.

2 Combinatorial Auctions as a Market Based Coordination Mechanism

Distributed decision making is considered an alternative to pure centralized scheduling systems as it facilitates the incorporation of local objectives, preferences and constraints of each resource in the decision making process. [6]. Combination of individual problem-solving and coordination/negotiation schemes is one of the research challenges in this area. [3] Market based allocation mechanisms are one of the most active branches of distributed task scheduling. It involves the creation of a production schedule based on the prices emerging from the bids sent by tasks. Each task proposes a bid trying to maximize its own objective. Each task has only access to its own information (objectives, preferences and constraints). The underlying idea is to allocate resources among the task by the creation of an ad-hoc market, setting prices through the search of equilibrium in an iterative process. There is no communication among agents representing tasks. Prices enable coordination of agents. Complex calculations are distributed among the participating agents, so that the problem is divided into several easier problems which can be solved in parallel. The communication overhead is low as it is limited to the exchange of bids and prices between agents and the market mechanism [7].

We propose a combinatorial auction mechanism to coordinate agents. In tasks scheduling problems tasks need a combination of time slots of multiple resources to do the operations. The use of combinatorial auctions as a negotiation and coordination mechanism in Multiagent Systems is appropriate in problems in which different valuations of interdependent items have to be considered. In combinatorial auction participants bid for a combination of different products. The valuation of an item depends on the combination of products it belongs to i.e. a bidder will make his valuation of an item based on which will be the other items that he will buy in the same auction. using iterative combinatorial auctions has many advantages. First, participants do not have to make bids over the set of all possible combinations of bids. Second, participants reveal in each iteration their private information and preferences. Third, iterative auctions are well-suited for dynamic environments (i.e. manufacturing environment) where participants and items get in and out in different moments [8].

3 Price-Setting Iterative Combinatorial Auction

Tasks scheduling problem can be modeled as an auction where time horizon is divided into slots that are sold in the auction. Tasks participate in the auctions as a bidders, trying to get the time slots of the resources that they need to perform the operation [9]. The mechanism will follow the main principles of distributed systems so the relevant information of the bidders (i.e. due date of the jobs, penalty for the delays) will be hidden to the rest of agents [10]. Prices show the preferences of other agents, and let the agent act consequently. There exists a central pool of resources (Resource-agent). Each resource has different abilities. The planning horizon of the resource is divided into time slots. These time slots are sold in an auction. There are a several tasks to be done to complete the jobs. The tasks need to get the necessary resources to be finished before the due date of the jobs they belong to. Tasks-agents bid for the time slot of the resources minimizing their cost function. An agent acts as a central node (Auctioneer-agent). Once the bids are sent the Auctioneer-agent will update the prices of the slots and the tasks-agents will remake their bids. This iterative process continues until prices are stabilized or a stop condition is fulfilled.

Job-agents are price-takers in the model. The price of a time slot (λ) is fixed by an iterative process. The prices of the slots are raised or lowered by a walrasian mechanism, e.g. an excess of demand raises prices and an excess of capacity lower them. There are many ways to update prices ($\lambda^{n+1} = \lambda^n + \Delta\lambda$) (e.g. constant increase or decrease of prices, proportional to the demand, proportional to the excess of demand).

4 Research Line

We search an updating price method for combinatorial auction that meet the needings of scheduling manufacturing systems in dynamic environments, e.g. they are intended to offer robustness, stability, adaptability, and efficient use of available resources through a modular and distributed design [5]. In our work we will use a specific problem as Job Shop scheduling problem, but we want to extend the conclusions that we will obtain to other problems of the same nature.

There is a similarity between iterative combinatorial auction and Lagrangian Relaxation Algorithm. The update of prices process is similar to the update of Lagrange multipliers iterative process. The updating methods used in the Lagrangian Relaxation Method can be used to update the prices in the iterative auction. [9]. Our objetive is to compare the different methods of updating prices based on those that update the lagrangian problem. [11] One of the most used method is subgradient methods [12, 13, 14]. There are also other methods derived from the former, e.g. surrogate gradient method [15, 16], conjugate subgradient method [11] or interleaved surrogate method [17, 18]. There is no work which compares the updating prices methods of combinatorial auctions for Job Shop problem to the best of our knowledge. We will define and implement a distributed task scheduling system based on the Job Shop Problem [1]. We want to compare the different methods of updating prices using as criteria convergence, stability and other points to study in a real case implementation as asynchronic computation. We will use different benchmark as we can find in [19] and the modifications suggested in [20].

Acknowledgements. This work has been partially supported Caja de Burgos through Projects 2009/00199/001 and 2009/00148/001.

References

1. Pinedo, E.P.M.L.: Scheduling, 2nd edn. Springer, New York (2008)
2. Shen, W.: Distributed manufacturing scheduling using intelligent agents. Intelligent Systems 17(1), 88–94 (2002)
3. Shen, W., et al.: Applications of agent-based systems in intelligent manufacturing: An updated review. Advanced Engineering Informatics 20(4), 415–431 (2006)
4. Lee, Kim: Multi-agent systems applications in manufacturing systems and supply chain management: a review paper. International Journal of Production Research 46(1), 233–265 (2008)
5. Ouelhadj, D., Petrovic, S.: A survey of dynamic scheduling in manufacturing systems. Journal of Scheduling 12(4), 417–431 (2009)
6. Kutanoglu, E., Wu, S.D.: On combinatorial auction and Lagrangean relaxation for distributed resource scheduling. IIE Transactions 31(9), 813–826 (1999)
7. Wellman, M.P., et al.: Auction Protocols for Decentralized Scheduling. Games and Economic Behavior 35(1-2), 271–303 (2001)
8. de Vries, S., Vohra, R.V.: Combinatorial Auctions: A Survey. Informs Journal on Computing 15(3), 284–309 (2003)
9. Dewan, P., Joshi, S.: Auction-based distributed scheduling in a dynamic job shop environment. International Journal of Production Research 40(5), 1173–1191 (2002)
10. Duffie, N.A.: Synthesis of heterarchical manufacturing systems. Comput. Ind. 14(1-3), 167–174 (1990)
11. Wang, J., et al.: An optimization-based algorithm for job shop scheduling. SADHANA 22, 241–256 (1997)
12. Geoffrion, A.M.: Lagrangean relaxation for integer programming. En Approaches to Integer Programming, 82–114 (1974), `http://dx.doi.org/10.1007/BFb0120690` (Accedido Mayo 20, 2009)
13. Fisher, M.L.: The Lagrangian Relaxation Method for Solving Integer Programming Problems. Management Science 50(suppl.12), 1861–1871 (2004)
14. Guignard, M.: Lagrangean relaxation. TOP 11(2), 151–200 (2003)
15. Camerini, P.M., Fratta, L., Maffioli, F.: On improving relaxation methods by modified gradient techniques. En Nondifferentiable Optimization, 26–34 (1975), `http://dx.doi.org/10.1007/BFb0120697` (Accedido Junio 3, 2009)
16. Brännlund, U.: A generalized subgradient method with relaxation step. Mathematical Programming 71(2), 207–219 (1995)
17. Zhao, X., Luh, P., Wang, J.: The surrogate gradient algorithm for Lagrangian relaxation method. In: Proceedings of the 36th IEEE Conference on Decision and Control, vol. 1, pp. 305–310 (1997)
18. Chen, H., Luh, P.: An alternative framework to Lagrangian relaxation approach for job shop scheduling. European Journal of Operational Research 149(3), 499–512 (2003)
19. Demirkol, E., Mehta, S., Uzsoy, R.: Benchmarks for shop scheduling problems. European Journal of Operational Research 109(1), 137–141 (1998)
20. Kreipl, S.: A large step random walk for minimizing total weighted tardiness in a job shop. Journal of Scheduling 3(3), 125–138 (2000)

A Software Tool for Harmonic Distortion Simulation Caused by Non-linear Household Loads

J. Baptista, R. Morais, A. Valente, S. Soares, J. Bulas-Cruz, and M.J.C.S. Reis

Abstract. In this paper we present a software tool to be used in residential/household generic power circuitry analysis and simulation, under non-linear loads. This tool can both be used by electrical engineers and by students of electrical engineering. It has an easy-to-use, friendly interface, and can be used to teach design techniques or as a laboratory tool to study the applicability of known methods to real world practical situations. Also, the users may supply their own data. The simulated results are very close to the measured ones.

Keywords: Residential power circuits, simulation, harmonic distortion, non-linear loads.

1 Introduction

Electric power generation, transportation and consumption are among the problems often faced by anyone working in the broad field of Electrical Engineering (EE). The problem of understanding and teaching harmonics mechanisms and its implications has attracted the scientific community for decades, and many are the works published in the field [1, 2, 3], to name only a few. For this reason, an introductory course focusing on electrical energy systems would be welcome in addition to the background of Electrical Engineering students, specially to those interested

---

J. Baptista · J. Bulas-Cruz
UTAD, Dept. Engenharias, Vila Real, Portugal
e-mail: {baptista, jcruz}@utad.pt

A. Valente · S. Soares · M.J.C.S. Reis
IEETA/UTAD
e-mail: {avalente, salblues, mcabral}@utad.pt

R. Morais
CITAB/UTAD
e-mail: rmorais@utad.pt

E. Corchado et al. (Eds.): SOCO 2010, AISC 73, pp. 31–38.

in electrical power generation, transportation, consumption, or electrical systems in general.

In this paper we present a computer application intended to be an easy-to-use tool to help the students and professionals of EE in solving problems while experimenting and comparing the available algorithms. The tool allows the users to try several loads on their own data, and then compare the results (and consequently infer the behaviour of the loads) in an easy-to-use environment that is similar to a laboratory. The computer application was named "HarmoSim", which stands for "Harmonic Simulation". This tool aims to simulate the behavior of residential/household power circuits with several kind of different non-linear loads. It allows the graphical visualization of different parameters related to circuit electric quality; among them are the total harmonic distortion, crest-factor, power-factor, and line-current. The harmonic spectrum and current and voltage waveforms are also plotted.

The next sections are dedicated to a small introduction to electrical energy quality, non-linear loads, and harmonic distortion parameters, in order to help to clarify the ideas that lead to the tool development.

2 Background

2.1 *Electrical Power Quality and Non-linear Loads*

The ongoing technological development demands for more efficient and better quality in the electrical energy supply. It is being viewed as an importante economic competitive factor to the industry, and services in general, particularly during the last decades. A great concern is being noticed, mainly in the industry, in order to minimize the economic risks resulting from a poor or bad Power Quality (PQ).

There are a great number of equipments very sensible to disturbances occurring at the line supply, particularly electronic equipments like computers, TV sets, among others. A huge effort is also being done in order to define criteria to evaluate PQ, and relating these criteria with the equipments' disturbances admissible functioning limits [4]. Also, there are a number of international standards limiting electric disturbances [5, 6].

A load is said to be non-linear when the consumed current and voltage waveforms have different shapes. Examples of non-linear loads are uninterruptible and switched power supplies, compact and traditional fluorescent lamps/tubes, controlled rectifiers. Figure 1(a) shows the current and voltage typical consumption waveforms for a personal computer (PC).

2.2 *Harmonic Distortion Indicators*

There are different ways to characterize and describe a periodic signal. Here we are interested in a set of parameters that may be used to measure and characterize the signals' harmonic distortion. The Fourier series is the tool used to analise the harmonic contents of a signal. Any periodic signal may be represented by its Fourier series, [7], as

$$y(t) = H_0 + \sum_{n=1}^{\infty} H_n^2 \sqrt{2} \sin(n\omega t + \varphi_n), \tag{1}$$

where H_n represents the n-th harmonic. H_0 corresponds to the DC component. The advantages and limitations of Fourier based analysis methods are widely documented in the literature in its various flavorous and applications [8, 9].

Electrical signals are odd, by its nature, having null DC (H_0) component and null even harmonics. For the most common harmonic producing devices the signals remain odd, which offers a further simplification for most power system studies. In fact, the presence of even harmonics is often a clue that there is something wrong (either with the load or with the measurement). Notable exceptions to this are the half-wave rectifiers and arc furnaces when the arc is random [4]. Also, loads like personal computers, electronic and magnetic ballasts and other electronic equipment, generate odd harmonics, [10, 11]. Usually, the higher-order harmonics (above 25th) are negligible for power system analysis [4]. Because we are working with typical household appliances, we will assumed only the first 25 odd harmonics.

From the set of techniques used to evaluate the voltage and current harmonic distortion the current root-mean-squared (RMS) value, power-factor, crest-factor, distortion power, frequency spectrum, and harmonic distortion are among the most traditionally used. We must know/understand these values in order to introduce possible correction actions.

The current RMS value can be defined by

$$I_{\text{RMS}} = \sqrt{\sum_{n=1}^{\infty} H_n^2},$$

which can be rewritten as

$$I_{\text{RMS}} = \sqrt{I_1^2 + I_3^2 + I_5^2 + \cdots + I_n^2}, \tag{2}$$

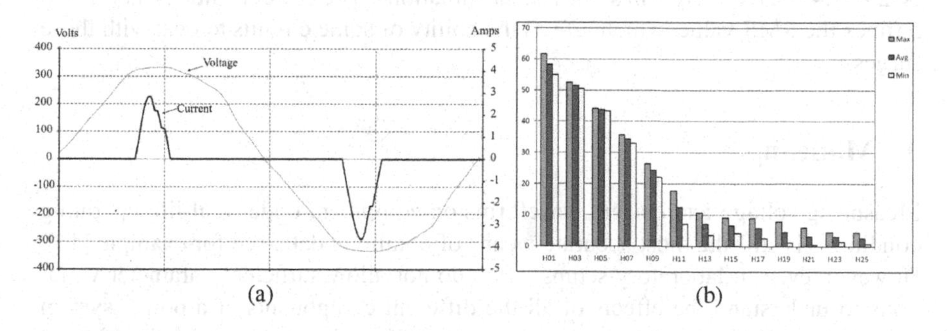

Fig. 1 Current and voltage typical consumption waveforms for a PC (a), and its measured harmonic current spectrum (b)

where I_n represents the current associated to the n-th odd harmonic. The greater the n the lower its influence in the I_{RMS} value. Usually the first harmonics will suffice to have a good approximation to the correct value. This value is also connected to the Joule's thermal effect.

According to IEEE 519-1992 recommended practices and requirements for harmonic control in electrical power systems [5], the Total Harmonic Distortion rate (THD) is defined as the ratio between the harmonics' RMS value and the fundamental component, i.e.,

$$\mathrm{THD}(\%) = \frac{\sqrt{\sum_{n=2}^{\infty} H_n^2}}{H_1} \times 100. \tag{3}$$

When no distortion is present, THD $= 0$; this means that a null or low THD is the goal to achieve.

The Power Factor, defined by

$$PF = \frac{\cos\varphi}{\sqrt{1+\mathrm{THD}_i^2}}, \tag{4}$$

where φ represents the phase-shift between the fundamental current and voltage components, is another parameter deeply affected by the introduction of harmonic distortion in the current: the far from one this parameter is, the greater the current distortion.

By definition, the Crest Factor, also called "peak-to-RMS-ratio" or "peak-to-average-ratio", corresponds to the ratio between the peak of the current or voltage value and its RMS value,

$$C_{\mathrm{V}} = \frac{V_{\max}}{V_{\mathrm{RMS}}}, \qquad C_{\mathrm{I}} = \frac{I_{\max}}{I_{\mathrm{RMS}}}. \tag{5}$$

If the signal is a perfect sinusoid (has no harmonic distortion) then the crest-factor value will be equal to $\sqrt{2}$. If the crest-factor value is different form $\sqrt{2}$ we have a distorted signal. Non-linear loads typically have current crest-factors above 1.5 ($\sqrt{2} \approx 1.414$), reaching 5 in some critical situations. The current value can be 1.5 to 5 times the RMS value, which affects the ability of some circuits to deal with these surges.

3 Modeling

Measuring voltage and current waveforms on equipment under real-life operating conditions is still the most accurate means of obtaining data, see for example [12]. However, even in laboratory setups, often do not allow sufficient parameter variations to understand the effects of all the different components of a power system on the system's behavior, and other methods and models may be used. See, for example, the introduction section of [13] for a synthetic review of several modeling methods. We will follow the time-domain modeling method proposed in [14].

In order to evaluate and understand the kind of disturbances that usually are injected in residential power circuitry during normal equipment operation/utilization, we have prepared a set of monitoring tests. During these tests we have measured the different parameters for different loads and functioning limits, using a PP-4300 Dranetz-BMI analyzer, (`www.dranetz-bmi.com`), and the data were treated in the DranView software package (`www.dranetz-bmi.com`). We have concluded that there are harmonic patterns repeating them selves for each kind of load, both for current and voltage waveforms. Figure 1(b) shows the spectrum harmonic results for the monitoring of a personal computer, one of the most used electronic loads in modern life. It produces harmonic current especially when there is a large concentration of them in a distribution system; it utilizes the switch mode power electronics technology which draws highly non-linear currents that contain large amounts of third and higher order harmonics [11]. For each current harmonic we have calculated the mean of each component value from all 20 tests. The amplitude and phase odd harmonics were then inserted in the tool's data base. The voltage case is not shown, but we have used an equivalent procedure.

From a harmonic point of view, the electric loads create uncertainty, because we do not know its nature (resistive, manly inductive or capacitive) as well as its operating point. The equivalent circuit of a non-linear load will always be a series and/or parallel of capacitors, resistors and inductors.

The common approach to characterize a certain non-linear load is to measure its harmonic voltage supply, $V(n)$, and the current it absorves, $I(n)$, being its impedance given by

$$Z(n) = \frac{V(n)}{I(n)}.$$

The standard harmonic pattern of all loads used during the development of the tool presented here was determined using this methodology.

An electrical installation may be approximated by a set of parallel impedances (loads) linked to the supply feeder. The total input current $I_i(n)$ will be equal to

$$I_i(n) = I_1(n) + I_2(n) + I_3(n) + \cdots + I_k(n),$$

where n represents the n-th harmonic and k the number of simultaneously connected loads.

To find the total voltage $V_i(n)$ it is necessary to determine the circuit equivalent impedance. Because all loads are parallel connected, the total impedance load will be given by

$$Z_{\text{eq}}(n) = \frac{1}{Y_{\text{eq}}(n)},$$

where

$$Y_{\text{eq}}(n) = Y_1(n) + Y_2(n) + Y_3(n) + \cdots + Y_k(n),$$

is the total admittance.

The harmonic voltage will then be given by the Ohm's law,

$$V_i(n) = Z_{eq}(n) \times I_i(n).$$

We are now able to estimate the current RMS value, equation 2, total harmonic distortion value (THD), equation 3, power-factor value (PF), equation 4, crest-factor value (CV), equation 5, the spectral harmonics (frequency), and the voltage and current (time) waveforms (recovered from the Fourier series, defined in equation 1).

4 Tool's Description and Simulation Example

The HarmoSim tool has been develop under the Visual Basic.NET programming environment. The tool has 6 main blocks/utilities: *loads management*, *loads analysis*, *circuitry simulation*, *reports history*, and *help*. Figure 2(a) shows the block diagram.

The *loads management* block is used to manage and characterize all loads existing in the data base (or to be added). The first 25 odd harmonics, phases, nominal current, and power voltage (according to measures) must be imputed. All these parameters associated with a particular load can be edited and/or deleted.

There are several characteristics associated with the *loads analysis* that can be viewed, edited or deleted, in the corresponding block. This includes the circuit harmonic spectrum, and current and voltage waveforms (both time and frequency domains). A Visual Basic.NET "ListView" is used to show: the harmonics, current and voltage in absolute value or as a percentage of the RMS value; the voltage and current spectrums plots; the time domain current and voltage plots; and the power-factor and crest-factor (both current and voltage). The total harmonic distortion, both for current and voltage, are calculated and presented according to [5] and [6] standards. A PDF report is generated, under user demand, that includes all the parameters and plots associated with the selected load.

The *circuitry simulation* block is where actually all the simulations and behaviour preview are done for single-phase or three-phase set of loads, choosed by the user. First, the user must choose the loads to simulate, from the ones existing in the data base; the different type, number of loads and their phases must be inserted. Next, the

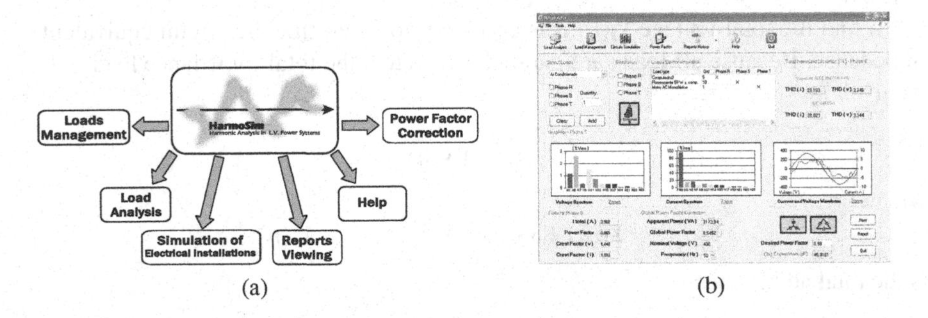

(a) (b)

Fig. 2 HarmoSim tool block diagram (a), and simulation block typical window (b)

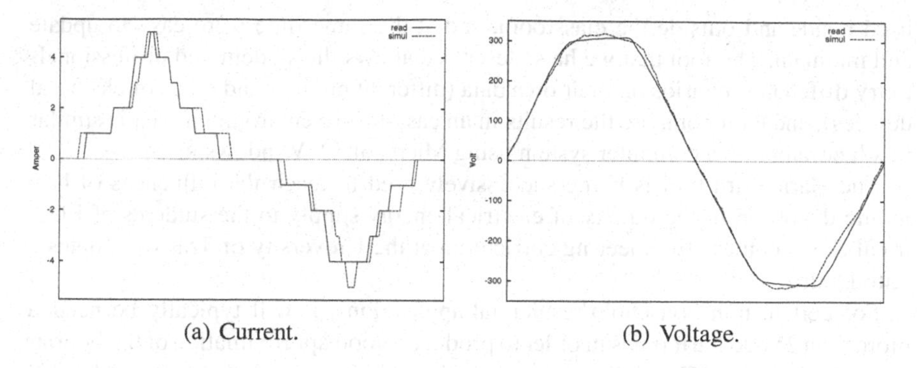

(a) Current. (b) Voltage.

Fig. 3 Results for 6 personal computers and 19, 58W, fluorescent tubes (OSRAM L 58W/765) without compensation

user chooses the phase to simulate (A, B, or C), because at the present tool's version only one phase at a time can be simulated. The different plots and parameters are then shown, including the current and voltage spectrum harmonics, total harmonic distortion, circuit power-factor, and crest-factors. Figure 2(b) shows the HarmoSim tool circuitry simulation window.

The *reports history* block may be used to see all the previously saved reports and look at a particular simulation. The *help* block explains, in a summarized way, the tool's functioning principles.

In order to compare the results of the simulations with the real values (measured using the Dranetz-BMI analyzer), we have prepared a set of monitoring tasks, using different loads in different operating conditions, that were implemented by our students. These monitoring tasks were conducted in the campus of the University of Trás-os-Montes e Alto Douro, houses and flats, and small industry. Here we present two of such case studies.

Figure 3 shows the current and voltage plots, measured and simulated, for 6 personal computers and 19, 58W, fluorescent lamps without capacitive compensation (OSRAM L 58W/765). The measured and simulated THD_V was 4.25% and 3.6%, respectively, and for THD_I these values were 33.1% and 31.16%. For the power-factor we had 0.95 and 0.95, and for the current crest-factor 1.9 and 2.14. The total current (A) 2.47 and 2.47.

As can be seen from the example above, good approximation results are achieved by the simulations produced by the HarmoSim tool. The average difference between the real (measured) and simulated signals are 4.5V and 1.3A. For the $l^\infty (||x||^\infty = \max |x_i|)$ error these differences are 22.2V and 5.3A.

5 Conclusions

Circuitry analysis and simulation tasks are among the problems that an electrical engineer often faces. Also, we experimented the need for tools that could be easily used inside and outside the classroom, and at the same time were easy to update and maintain. The tool that we have described allows the student and professionals to try different scenarios on their own data (different number and types of electrical devices), and then compare the results in an easy-to-use environment that is similar to a laboratory, in a computer system using Microsoft™ Windows®.

The HarmoSim tool is being successively used to teach the influences of harmonic distortion in the quality of electrical energy supply to the students of Electrical & Computers Engineering curriculum at the University of Trás-os-Montes e Alto Douro.

For certain non-household/residential applications, it will typically be needed more than 25 odd harmonics in order to produce a good approximation of the Fourier series to the signal. The authors are currently working on a new version of the tool in order to include a greater number of harmonics (> 25), and all the three-phases during the simulation process at once.

References

1. Schlabbach, J.: Proceedings of the 12th IEEE Mediterranean, Electrotechnical Conference, MELECON 2004, vol. 3, pp. 1153–1156 (2004)
2. Rob, R.A., Jewell, W.T.: Industrial and Commercial Power Systems Technical Conference, pp. 70–77 (1993)
3. Pak, L.F., Dinavahi, V., Chang, G., Steurer, M., Ribeiro, P.F.: IEEE Transactions on Power Delivery 22, 1218 (2007)
4. Dugan, R.C., McGranaghan, M.F., Beaty, H.W.: Electrical Power Systems Quality. McGraw-Hill, New York (1996)
5. I. 519-1992. Recommended practices and requirements for harmonic control in electrical power systems (1992)
6. I. 60555-1. Disturbances in supply systems caused by household appliances and similar electrical equipment—Parte 1: Definitions (1997)
7. Brigham, E.O.: The Fast Fourier Transform. Prentice-Hall, Englewood Cliffs (1974)
8. van Vleck, E.B.: Science 39, 113 (1914)
9. Cooley, J.W.: IEEE Signal Processing Magazine 9(1), 10 (1992)
10. Arrillaga, J., Watson, N.R.: Power System Harmonics. John Wiley and Sons, London (1985)
11. Venkatesh, C., Kumar, D.S., Sarma, D.S., Sydulu, M.: Fifteenth National Power Systems Conference (NPSC), IIT Bombay, pp. 592–597 (2008)
12. Steurer, M., Woodruff, S.: In: IEEE Power Engineering Society General Meeting, vol. 1, pp. 773–776 (2004)
13. Soliman, S.A., Alammari, R.A.: Electric Power Systems Research 72(2), 147 (2004)
14. Girgis, A.A., Quaintance III, W.H., Qiu, J., Makram, E.B.: A Time-domain Three-phase Power System Impedance Modeling Approach for Harmonic Filter Analysis. IEEE Transactions on Power Delivery 8(2), 504 (1993)

A Multiobjective Variable Neighborhood Search for Solving the Motif Discovery Problem

David L. González-Álvarez, Miguel A. Vega-Rodríguez, Juan A. Gómez-Pulido, and Juan M. Sánchez-Pérez

Abstract. In this work we approach the Motif Discovery Problem (MDP) by using a trajectory–based heuristic. Identifying common patterns, motifs, in deoxyribonucleic acid (DNA) sequences is a major problem in bioinformatics, and it has not yet been resolved in an efficient manner. The MDP aims to discover patterns that maximize three objectives: support, motif length, and similarity. Therefore, the use of multiobjective evolutionary techniques can be a good tool to get quality solutions. We have developed a multiobjective version of the Variable Neighborhood Search (MO–VNS) in order to handle this problem. After accurately tuning this algorithm, we also have implemented its variant Multiobjective Skewed Variable Neighborhood Search (MO–SVNS) to analyze which version achieves more complete solutions. Moreover, in this work, we incorporate the hypervolume indicator, allowing future comparisons of other authors. As we will see, our algorithm achieves very good solutions, surpassing other proposals.

1 Introduction

The application of evolutionary algorithms for solving optimization problems has been intense in recent decades. An evolutionary algorithm uses some mechanisms inspired by biological evolution: reproduction, mutation, recombination, and selection. The solutions of an optimization problem play the role of individuals in a population, and a fitness function determines the quality of them. All these solutions evolve after the repeated application of the above functions. There are many types of evolutionary algorithms; in this paper we will focus our attention on trajectory–based heuristics. These techniques begin with a single solution that is updated by exploring the neighborhood, forming a trajectory. The search ends when we reach a

David L. González-Álvarez · Miguel A. Vega-Rodríguez ·
Juan A. Gómez-Pulido · Juan M. Sánchez-Pérez
University of Extremadura, Polytechnic School, Cáceres, Spain
e-mail: {dlga,mavega,jangomez,sanperez}@unex.es

E. Corchado et al. (Eds.): SOCO 2010, AISC 73, pp. 39–46.

maximum number of iterations, we find a solution with acceptable quality, or when we detect a lack of progress. An example of trajectory–based algorithm is the Variable Neighborhood Search (VNS), algorithm applied in this work.

The main goal of bioinformatics is to understand biological processes. To increase our knowledge in this regard, numerous investigations are underway in areas such as the sequence alignment, gene finding, genome assembling, alignment of protein structures, protein structure prediction, and protein–protein interactions. In this paper we focus on finding patterns in deoxyribonucleic acid (DNA) sequences, the Motif Discovery Problem (MDP). The motif discovery is a fundamental problem applied to the specific task of discovering novel transcription factor binding sites in DNA sequences [1]. The MDP is an NP–complete optimization problem, so that, the use of evolutionary techniques can be a good tool to find quality patterns, motifs. This problem aims to maximize three conflicting objectives, so we should use a multiobjective algorithm. To solve this problem we apply a multiobjective version of the VNS algorithm, the Multiobjective Variable Neighborhood Search (MO–VNS). In addition we also analyze the performance of a skewed version of MO–VNS, the Multiobjective Skewed Variable Neighborhood Search (MO–SVNS). The results obtained by our heuristics improve other well known methods of finding motifs such as AlignACE, MEME, and Weeder, as well as achieve better results than other researchers in the field.

This paper is organized as follows. In the next section we describe the motif discovery problem. Section 3 details the adjustments and modifications made on the evolutionary algorithm used to find motifs. Section 4 shows the results obtained with our algorithms. In Section 5 we perform different comparisons with other authors. Finally, in Section 6, we show the conclusions and future work.

2 Motif Discovery Problem

Motifs are recurring patterns of short sequences of DNA, RNA, or proteins that usually serves as a recognition site or active site. The same motif can be found in a variety of types of organisms. Motifs are usually very short (up to 30 nucleotides) and gapless, discovering motifs in the midst of all the biological information in DNA is not an easy task. The Motif Discovery Problem (MDP) finds patterns overrepresented among all this biological information. We align the sequences of a collection of regulatory regions of genes that are believed to be coregulated to get motif instances.

Recent publications have proposed a new method for discovering motifs using multiobjective approach [2]. This method maximizes three conflicting objectives: support, motif length, and similarity: *Support*: Is the number of sequences used to form the motif. If we do not find some candidate motif, this sequence will not be taken into account. *Motif length*: Is the number of nucleotides that compose the motif. A high motif length gives confidence that the solution has a real biological significance. *Similarity*: This objective maximizes the similarity of subsequences that make up the resulting motif. First we generate a position weight matrix from the motif patterns found, then we calculate the dominance value (*dv*) of every nucleotide at each motif position. We select the highest value of each motif position using the following expression:

$$dv(i) = max_b\{f(b,i)\}i = 1,...,l \tag{1}$$

where *f(b,i)* is the score of nucleotide *b* in column *i* in the position weight matrix, *dv(i)* is the dominant nucleotide (higher dominance value) in column *i*, and *l* is the motif length. Then, we define the similarity value of a motif as the average of all the dominance value for all position weight matrix columns. As is indicated in the following expression:

$$Similarity(M) = \frac{\sum_{i=1}^{l} dv(i)}{l} \tag{2}$$

3 Multiobjective Variable Neighborhood Search

The Variable Neighborhood Search (VNS) is a heuristics for solving optimization problems created by Pierre Hansen and Nenad Mladenovic [3], whose basic idea is systematic change of neighborhood within a local search.

Algorithm 1. Pseudocode for MO–VNS

```
 1:  paretoSolutions <- 0
 2:  for ( i = 1 to MAXGENERATIONS ) do
 3:    S <- GenerateInitialSolution {randomly}
 4:    if notDominatedInParetoSolutions(S) then
 5:      /* Add solution and remove solutions dominated by S */
 6:    end if
 7:    k <- 1 //we initialize the neighborhood environment
 8:    while k < kmax do
 9:     S2 <- MutationAndLocalSearchFunctions(S,k)
10:      if notDominatedInParetoSolutions (S2) then
11:        /* Add solution and remove solutions dominated by S2 */
12:      end if
13:      if S2 dominates S then
14:        S <- S2
15:        k <- 1
16:      else
17:        k <- k + 1 //we increase the neighborhood environment
18:      end if
19:    end while
20: end for
```

To solve the MDP we have used a multiobjective version of VNS algorithm, named Multiobjective Variable Neighborhood Search (MO–VNS). This new algorithm is based on the classic VNS to be applicable in multiobjective problems. An outline of the algorithm is shown in Algorithm 1. Firstly, we initialize the Pareto front (line 1 of Algorithm 1). Then we create a random solution that will improve during the evolution progress. After evaluating this individual, we check if it is not dominated by any of the solutions currently included in the Pareto front, if so, we will add this new solution to the Pareto front and we will remove all solutions which are dominated by it (line 5). Now it is when we examine the neighbors of our individual applying mutation and local search functions (lines 8 to 18). If we do not improve our solution, we will intensify the mutation and local search functions (examining more distant neighbors).

After accurately tuning MO–VNS, we also have implemented its variant Multiobjective Skewed Variable Neighborhood Search (MO–SVNS). In MO–SVNS we do not always take the best solution as a reference for further evolving, we will use slightly worse solutions to try to overcome possible local maximum. For this, we use the *alpha* parameter (range [0.0, 1.0]) and we define a *distance* function to specify the loss quality rate allowed. We use a tridimensional *distance* function that is calculated by using the difference between the best cost value from solutions *S* and *S2* and the worst cost from all the Pareto front solutions, for each objective (Equation 3). The pseudocode for MO–SVNS is the same as the MO–VNS except in line 13 that is shown in Algorithm 2.

$$Distance(S2,S)_{objn} = |Best_{objn}(S2,S) - Worst_{objn}(ParetoFront)| \quad (3)$$

Algorithm 2. Pseudocode for MO–SVNS

```
13:      if S2 dominates ( S + ALPHA * Distance(S2,S) ) then
```

In both algorithms the individuals include the necessary information to form a possible motif. An individual represents the starting location (*si*) of the potential motif on all the sequences. As we work with data sets with different number of sequences, the definition of individuals should be adapted to each data set. Our definition of the individual also includes the motif length. In Fig. 1 we see the representation of an individual in our algorithms.

	Seq. 1	Seq. 2	Seq. 3		Seq. n
Length	s_1	s_2	s_3	...	s_n

Fig. 1 Representation of an individual

4 Experimental Results

In this section we explain the experiments performed to obtain the best configuration of MO–VNS. For each experiment we have performed 30 independent runs to assure its statistical relevance. The results are measured using the Hypervolume indicator [4], and the reference volume is calculated using the maximum values of each objective in each data set. We used twelve real sequence data sets (see Table 1) corresponding to alive beings (fly –dm–, human –hm–, mouse –mus–, and yeast –yst–), as a benchmark for the discovery of transcription factor binding sites [5], which were selected from TRANSFAC database [6].

Table 1 Data sets properties

Data set	Sequences	Size
dm01r	4	1500
dm04r	4	2000
dm05r	5	2500
hm03r	10	1500
hm04r	13	2000
hm16r	7	3000
mus02r	9	1000
mus07r	4	1500
mus11r	12	500
yst03r	8	500
yst04r	7	1000
yst08r	11	1000

Table 2 Experiment 1: LS depth

Data set	30	40	60	120	300
dm01r	0.762	**0.766**	0.764	0.763	0.759
dm04r	**0.772**	0.768	0.760	0.766	0.764
dm05r	0.796	0.794	**0.801**	0.794	0.791
hm03r	0.581	0.582	0.595	0.601	**0.620**
hm04r	0.469	0.488	0.492	0.513	**0.539**
hm16r	0.692	0.680	0.692	0.703	**0.707**
mus02r	0.592	0.602	0.621	0.619	**0.622**
mus07r	0.747	**0.751**	0.751	0.749	0.744
mus11r	0.521	0.519	0.537	0.551	**0.572**
yst03r	0.654	0.646	0.649	**0.658**	0.655
yst04r	0.692	0.689	**0.701**	0.690	0.700
yst08r	0.624	0.632	0.638	**0.640**	0.632

In all our experiments we have distinguished nine levels of neighborhood, using kmax=9. The first experiment aims to select the best *Local Search (LS) Depth*. This experiment has been performed using the following values: 30, 40, 60, 120, and 300, where these values indicate the number of neighbors evaluated in the local search. Analyzing the results obtained, as shown in Table 2, we conclude that the best results in most of the data sets are achieved using the value 300. We also performed experiments with bigger values, but we obtained worse results. With the second experiment, we select the best *Mutation Shift* (the magnitude of the effect of a mutation on the individual) for all data sets. This experiment was performed with the values: 1%, 5%, 10%, 15%, and 20%. Analyzing the results obtained we get the best results with a *Mutation Shift* of 20%. So, in order to obtain more accurate conclusions, we performed more runs with the values: 25%, 30%, and 35%. Following this, we see in Table 3 how the best results were obtained with the value 30%. Finally, with the last experiment we compare the MO–VNS with its skewed version, the MO–SVNS. For this, we have performed runs with different values of *Alpha*: 0.0, 0.2, 0.4, 0.6, 0.8, and 1.0, as we can see in Table 4. The best results are obtained with *Alpha*=0.0, that is, with the MO–VNS.

Table 3 Experiment 2: Mutation shift

Data set	1%	5%	10%	15%	20%	25%	30%	35%
dm01r	0.726	0.759	0.768	0.765	0.772	**0.779**	0.774	0.767
dm04r	0.725	0.764	0.772	0.775	0.773	0.784	**0.788**	0.779
dm05r	0.773	0.791	0.811	0.809	0.807	**0.811**	0.810	0.807
hm03r	0.558	0.620	0.632	0.648	0.649	**0.662**	0.661	0.650
hm04r	0.485	0.539	0.551	0.548	0.552	**0.571**	0.569	0.569
hm16r	0.650	0.707	0.737	0.738	0.743	0.779	**0.790**	0.758
mus02r	0.574	0.622	0.643	0.652	0.643	0.658	**0.661**	0.657
mus07r	0.712	0.744	0.750	0.760	0.760	0.761	**0.764**	0.762
mus11r	0.491	0.572	0.584	0.597	0.595	0.609	**0.616**	0.615
yst03r	0.607	0.655	0.673	0.680	0.685	0.691	**0.693**	0.689
yst04r	0.637	0.700	0.704	0.723	0.726	0.740	0.745	**0.748**
yst08r	0.559	0.632	0.669	0.678	0.688	0.709	**0.720**	0.704

Table 4 Experiment 3: Alpha

Data set	0.0	0.2	0.4	0.6	0.8	1.0
dm01r	0.774	0.771	0.769	0.774	**0.775**	0.773
dm04r	**0.788**	0.778	0.783	0.781	0.774	0.777
dm05r	0.810	0.807	0.806	0.807	0.803	**0.818**
hm03r	**0.661**	0.657	0.658	0.652	0.659	0.648
hm04r	**0.569**	0.556	0.554	0.549	0.568	0.559
hm16r	**0.790**	0.769	0.763	0.742	0.766	0.748
mus02r	**0.661**	0.653	0.652	0.646	0.651	0.654
mus07r	**0.764**	0.758	0.761	0.763	0.761	0.763
mus11r	0.616	0.610	0.614	0.614	0.602	**0.620**
yst03r	**0.693**	0.693	0.688	0.687	0.688	0.682
yst04r	**0.745**	0.743	0.738	0.738	0.740	0.733
yst08r	**0.720**	0.698	0.713	0.706	0.700	0.700

5 Comparison with Other Authors

In this section we compare our algorithm with other motif discovery algorithms as MOGAMOD algorithm [7], and with other well–known methods such as AlignACE [8], MEME [9], and Weeder [10]. We also analyze the differences between the run-times of each method, as in any optimization problem we must be able to find good solutions in a reasonable time. We concentrate our comparisons on yst04r, yst08r, and hm03r sequence data sets.

Firstly, we compare our results with those obtained by the MOGAMOD algorithm, other multiobjective algorithm dedicated to discover motifs. To demonstrate the superiority of our algorithm we have performed comparisons in two ways: first we compare the similarities of each algorithm keeping constant the other two objectives: support and motif length (Table 5: Similarity comparisons). Then, we compare the motif lengths obtained by each algorithm keeping constant the other two objectives, in this case: support and similarity (Table 5: Length comparisons). In both comparisons we can see how the MO–VNS algorithm achieves better solutions. All these comparisons have been performed with the two highest values of support in each data set. In addition to compare the results with MOGAMOD, we have compared our solutions with other well–known methods as AlignACE, MEME, and Weeder. Table 6 gives the results of this comparison. We can see how MO–VNS achieves larger solutions than the other methods, maintaining high values in each objective. Although the MOGAMOD algorithm achieves solutions with high similarities, these solutions have a low support. Our algorithm finds longer motifs that have high similarity values with maximum values of support. As we can see in Table 6, the solutions always maintain a balance between the values of the three objectives. We see how as the support and the motif length values increase, the similarity value decreases. However, with the same value of support, as the motif length decreases, the similarity value raises. In Table 7 we show our best results for the three data sets used.

Table 5 Similarity and length comparisons

		Similarity comparisons			Length comparisons		
		MO–VNS		MOGAMOD	MO–VNS	MOGAMOD	
	Support	Length	Similarity	Similarity	Length	Length	Similarity
yst04r	6	14	0.84	0.77	22	14	0.77
		13	0.85	0.81	15	13	0.81
	7	9	0.90	0.80	18	9	0.80
		8	0.91	0.84	15	8	0.84
yst08r	10	12	0.83	0.79	15	12	0.79
		11	0.86	0.82	13	11	0.82
	11	11	0.86	0.77	16	11	0.77
		10	0.88	0.80	13	10	0.80
hm03r	9	13	0.78	0.77	14	13	0.77
		11	0.79	0.78	13	11	0.78
	10	11	0.79	0.74	14	11	0.74
		10	0.79	0.79	10	10	0.79
		9	0.81	0.81	9	9	0.81

Table 6 Comparison of the predicted motifs by five methods

Data set	Method	Support	Length	Similarity	Predicted motif
yst04r	AlignACE	N/A	10	N/A	CGGGATTCCA
	MEME	N/A	11	N/A	CGGGATTCCCC
	Weeder	N/A	10	N/A	TTTTCTGGCA
	MOGAMOD	5	14	0.84	CGAGCTTCCACTAA
		6	14	0.77	CGGGATTCCTCTAT
	MO-VNS	6	22	0.77	TTTTTTCCTTCTTTCTTTTTTT
		7	18	0.80	AATAAAAAAAAAAAAAAA
yst08r	AlignACE	N/A	12	N/A	TGATTGCACTGA
	MEME	N/A	11	N/A	CACCCAGACAC
	Weeder	N/A	10	N/A	ACACCCAGAC
	MOGAMOD	7	15	0.84	TCTGGCATCCAGTTT
		7	15	0.87	GCGACTGGGTGCCTG
	MO-VNS	10	23	0.72	TTTTTTCTTTTTTATTTTTTTTT
		11	20	0.75	TTTTTTTTTATTTTTTTTTT
hm03r	AlignACE	N/A	13	N/A	TGTGGATAAAAAA
	MEME	N/A	20	N/A	AGTGTAGATAAAAGAAAAAC
	Weeder	N/A	10	N/A	TGATCACTGG
	MOGAMOD	7	22	0.74	TATCATCCCTGCCTAGACACAA
		7	18	0.82	TGACTCTGTCCCTAGTCT
	MO-VNS	9	18	0.73	TTGAAAAAACAAAAAAAA
		10	14	0.74	AAAAAAGCAAAAAA

The solutions shown improve MOGAMOD solutions in at least one objective. Finally, to complete our comparisons we compare the time requirements of the four methods (the runtime of Weeder is not available). MO–VNS experiments have been performed on a Pentium IV 2.33 GHz CPU with 1 GB of memory, and the rest of experiments have been run on Pentium IV 3.0 GHz CPU with 1 GB of memory.

Fig. 2 shows the results of this comparison, we see as the runtimes obtained by MO–VNS and MOGAMOD are very similar, taking into account the differences between the machines where the experiments have been performed. In summary, the MO–VNS algorithm discovers better motifs than other methods for finding patterns achieving similar runtimes to the best of them.

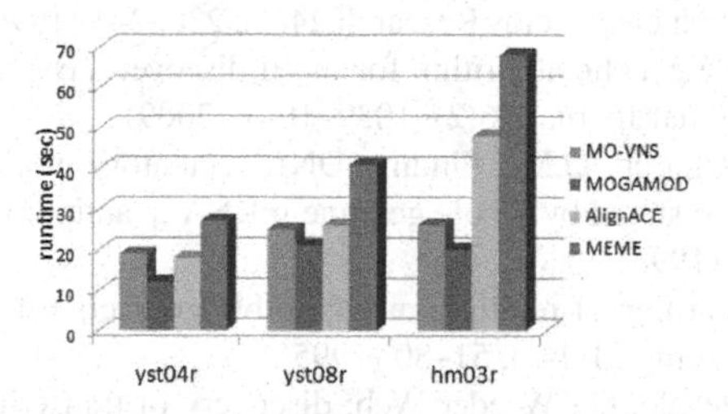

Fig. 2 Comparison of runtimes by four methods

Table 7 Best solutions of MO–VNS

	Support	Length	Similarity
yst04r	7	10	0.900
	7	11	0.896
	7	12	0.892
	7	13	0.857
	7	14	0.847
	7	15	0.840
yst08r	11	12	0.811
	11	13	0.800
hm03r	10	11	0.790

6 Conclusions and Future Work

In this work we have proposed the use of a trajectory–based algorithm to discover optimal motifs, the MO–VNS. We have adjusted all the parameters to obtain the best configuration of the algorithm for twelve data sets of different properties. Comparing the performance of MO–VNS with other motif discovery techniques such as MOGAMOD, AlignACE, MEME, or Weeder, we may say that it performs well because it improves the results obtained by those.

As future work we have the intention of implementing the two standard algorithms in multiobjective optimization: NSGA–II and SPEA2, comparing their results with the ones accomplished by the MO–VNS algorithm. We also pretend to apply some parallel technique to achieve good solutions in very large data sets.

Acknowledgements. This work was partially funded by the Spanish Ministry of Science and Innovation and ERDF (the European Regional Development Fund), under the contract TIN2008–06491–C04–04 (the M* project). David L. González–Álvarez is supported by a research grant from Fundación Valhondo (Spain).

References

1. Stine, M., Dasgupta, D., Mukatira, S.: Motif discovery in upstream sequences of coordinately expressed genes. In: The 2003 Congress on Evolutionary Computation (CEC 2003), December 2003, vol. 3, pp. 1596–1603 (2003)
2. Kaya, M.: Motif discovery using multi-objective genetic algorithm in biosequences. In: Berthold, M., Shawe-Taylor, J., Lavrač, N. (eds.) IDA 2007. LNCS, vol. 4723, pp. 320–331. Springer, Heidelberg (2007)
3. Mladenovic, N., Hansen, P.: Variable neighborhood search. Computers and Operations Research 24, 1097–1100 (1997)
4. Fonseca, C.M., Paquete, L., Lopez–Ibanez, M.: An improved dimension–sweep algorithm for the hypervolume indicator. In: IEEE Congress on Evolutionary Computation (CEC 2006), July 2006, pp. 1157–1163 (2006)
5. Tompa, M., et al.: Assessing computational tools for the discovery of transcription factor binding sites. Nature Biotechnology 23(1), 137–144 (2005)
6. Wingender, E., Dietze, P., Karas, H., Knüppel, R.: TRANSFAC: a database on transcription factors and their DNA binding sites. Nucleic Acids Research 24(1), 238–241 (1996)
7. Kaya, M.: MOGAMOD: Multi–objective genetic algorithm for motif discovery. Expert Systems with Applications: An International Journal 36(2), 1039–1047 (2009)
8. Roth, F.P., Hughes, J.D., Estep, P.W., Church, G.M.: Finding DNA regulatory motifs within unaligned noncoding sequences clustered by whole genome mRNA quantitation. Nature Biotechnology 16(10), 939–945 (1998)
9. Bailey, T.L., Elkan, C.: Unsupervised learning of multiple motifs in biopolymers using expectation maximization. Machine Learning 21(1-2), 51–80 (1995)
10. Pavesi, G., Mereghetti, P., Mauri, G., Pesole, G.: Weeder Web: discovery of transcription factor binding sites in a set of sequences from co–regulared genes. Nucleic Acids Research 32, 199–203 (2004)

Solving the Routing and Wavelength Assignment Problem in WDM Networks by Using a Multiobjective Variable Neighborhood Search Algorithm

Álvaro Rubio-Largo, Miguel A. Vega-Rodríguez, Juan A. Gómez-Pulido, and Juan M. Sánchez-Pérez

Abstract. At the present time, the future of communications is focused on optical fiber. The most promising technology is based on Wavelength Division Multiplexing (WDM). This technique divides the bandwidth into different wavelengths avoiding possible bottlenecks, therefore it takes full advantage of the bandwidth of the optical networks. A problem comes up when it is necessary to accomplish a set of transmission demands. This is known as Routing and Wavelength Assignment problem (RWA problem). There are two different types: Static-RWA (unicast demands, the most usual ones) and Dynamic-RWA (multicast demands). In this paper we have focused on the first type, Static-RWA. To solve it, we have used a multiobjective evolutionary algorithm. We have chosen the Variable Neighborhood Search algorithm (VNS), but in a multiobjective context (MO-VNS). After an exhaustive comparison with other authors, we conclude that this algorithm obtains much better results than their approaches.

1 Introduction

The technology based on optical fiber enables transmitting information from one device to another by sending pulses of light through an optical fiber. This technology has revolutionized the telecommunications industry.

The most promising technology to take full advantage of the bandwidth of the optical networks is based on Wavelength Division Multiplexing (WDM). This technique multiplies the available capacity of an optical fiber by adding new channels, each channel on a new wavelength of light. This approach ensures fluent communications between different devices interconnected by a specific optical network. Therefore these devices are able to send and receive information without

Álvaro Rubio-Largo · Miguel A. Vega-Rodríguez ·
Juan A. Gómez-Pulido · Juan M. Sánchez-Pérez
University of Extremadura, Polytechnic School, Cáceres, Spain
e-mail: {arl,mavega,jangomez,sanperez}@unex.es

E. Corchado et al. (Eds.): SOCO 2010, AISC 73, pp. 47–54.
springerlink.com

bottlenecks [1]. When it is necessary to interconnect a set of source-destination paths a problem comes up, this problem is known as Routing and Wavelength Assignment (RWA). In this problem is required to fulfill the next two constraints [2]: Wavelength conflict constraint and Wavelength continuity constraint.

In RWA problem there are two different varieties: Static-RWA and Dynamic-RWA. Both varieties are NP-complete problems, so it is very common the use of heuristics to solve them in literature [3].

Nowadays most of the WANs (Wide Area Networks) are oriented to precontracted traffic services, so in this paper we have decided to implement an evolutionary algorithm to solve the Static-RWA problem (the most usual one). The Variable Neighborhood Search algorithm (VNS) [4] is our approach, but it was adapted to this Multiobjective Optimization Problem (MOP), we refer to it as MO-VNS. We have adjusted this algorithm using two real-world topologies. The first one from USA (NSF), and the second one from Japan (NTT). As well, we have compared with other authors and finally we conclude that the MO-VNS algorithm obtains much better results than the other proposed approaches.

The rest of this paper is organized as follows. In Section 2 we present in a formal way the RWA problem. A description of the MO-VNS algorithm appears in Section 3. In Section 4 we present the test instances used, the experimental results and a comparison with other approaches. Finally, the conclusions and future work are left for Section 5.

2 RWA Problem

This Section shows the problem formulation and the objective functions of the problem, that is to say, we present in a formal way the RWA problem.

In this paper, an optical network is modeled as a direct graph $G = (V, E, C)$, where V is the set of nodes, E is the set of links between nodes and C is the set of available wavelengths for each optical link in E.

$(i, j) \in E$: Optical link from node i to node j; $i, j \in V$
$c_{ij} \in C$: Number of channels or different wavelengths at link (i, j)
$u = (s, d)$: Unicast request u with source node s and a destination node d, where $s, d \in V$
U : Set of unicast request, where $U = \{ u \mid u \text{ is an unicast request} \}$
$u^{\lambda}_{i,j}$: Wavelength λ assigned to the unicast request u at link (i, j)
l_u : Lightpath or set of links between a source node s_u and destination node d_u; with the corresponding wavelength assignment in each link (i, j)
L_u : Solution of the RWA problem considering the set of U requests.

Notice that $L_u = \{l_u | l_u$ is the set of links with their corresponding wavelength assignment $\}$. Using the above definitions, the RWA problem may be stated as a Multiobjective Optimization Problem (MOP) [5], searching the best solution L_u that simultaneously minimizes the following two objective functions:

1. Hop Count:

$$y1 = \sum_{u \in U} \sum_{(i,j) \in l_u} \Phi_j \ where \begin{Bmatrix} \Phi_j = 1 \ \text{if} \ (i,j) \in l_u \\ \Phi_j = 0 \ \text{if} \ otherwise \end{Bmatrix} \quad (1)$$

2. Number of wavelength (λ) conversions:

$$y2 = \sum_{u \in U} \sum_{i \in V} \varphi_j \ where \begin{Bmatrix} \varphi_j = 1 \ \text{if} \ i \in V switches \ \lambda \\ \varphi_j = 0 \ \text{if} \quad otherwise \end{Bmatrix} \quad (2)$$

Furthermore, we have to fulfill *the wavelength conflict constraint*: Two different unicast transmissions must be allocated with different wavelengths when they are transmitted through the same optical link (i, j).

3 Multiobjective Variable Neighborhood Search Algorithm

The Variable Neighborhood Search (VNS) algorithm is a trajectory-based algorithm created by P. Hansen and N. Mladenovic [4]. It starts from an initial individual, and performs many changes of neighborhood within a local search. When the search does not move forward, the algorithm increases the size of the neighborhood (k).

As the Static-RWA problem is a MOP, we had to adapt the VNS algorithm to a multiobjective context (MO-VNS). In MOP it is necessary to compare candidate solutions, the most common way to compare them is known as Pareto dominance. In a formal way [6], we could define the Pareto set as follows:

If a MOP has n different functions to be minimized $(f_1(x), f_2(x), ..., f_n(x))$. A solution x_1 dominates a solution x_2 if:

$$\begin{gathered} \forall i \in \{1,2,...,n\} : f_i(x_1) \leq f_i(x_2) \\ \wedge \\ \exists j \in \{1,2,...,n\} : f_j(x_1) < f_j(x_2) \end{gathered} \quad (3)$$

The individual in this paper was designed as it is shown in Fig. 1. We have developed a modified version of Yen´s algorithm [7]. We modified it by introducing an

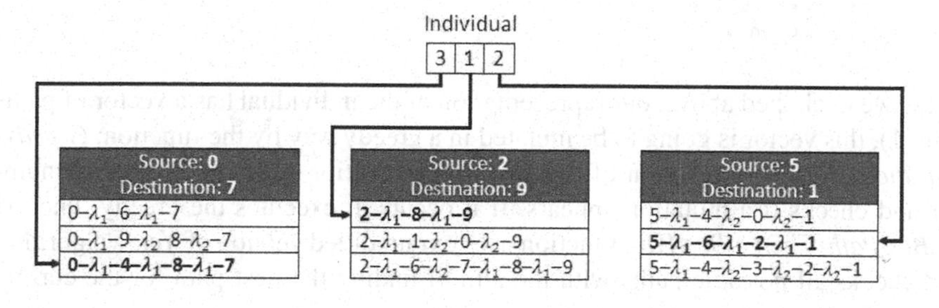

Fig. 1 Structure of an Individual

heuristic to assign the wavelengths. Whenever possible we try to assign always the same wavelength that was previously assigned (previous hop), but if not or it is the first assignation, we assign the first free wavelength. Using this modified version, we create a list of possible routes (including wavelengths) and we select one. This selection is stored in a vector. We repeat this process for every lightpath.

We have implemented another variant of the VNS, named Skewed Variable Neighborhood Search (SVNS), also in a multiobjective context (MO-SVNS). Both are very similar, the main difference between them is that the MO-SVNS is more receptive in accepting a new solution. To understand how they work, in Algorithm 1 we present an explanatory pseudocode. The *distance* function (line 15) is a function that quantifies the distance between *S* and *S'*. This function calculates the distance for each objective function as it is shown in the following equation:

$$\begin{aligned} Distance_{y_1} &= |Best_{y_1}(S,S') - Worst_{y_1}(ParetoFront)| \\ Distance_{y_2} &= |Best_{y_2}(S,S') - Worst_{y_2}(ParetoFront)| \end{aligned} \quad (4)$$

The distance function is only used by MO-SVNS, because the parameter α takes values between 0 and 1. MO-VNS does not use this distance function because α is always 0.

Algorithm 1. MO-VNS ($\alpha = 0$) and MO-SVNS ($0 < \alpha \leq 1$)

```
 1:  ParetoFront <- 0
 2:  while not time-limit do
 3:     S <- GenerateIndividual( )
 4:     if IsNotDominatedByParetoFrontSolutions(S) then
 5:           ParetoFront <- AddIndividual(S)
 6:           ParetoFront <- RemoveIndividualsDominatedBy(S)
 7:     end if
 8:     k <- 1
 9:     while k < kmax do
10:          S' <- GreedyMutation(k, S)
11:          if IsNotDominatedByParetoFrontSolutions(S') then
12:              ParetoFront <- AddIndividual(S')
13:              ParetoFront <- RemoveIndividualsDominatedBy(S')
14:          end if
15:          if S' dominates (S + alpha*distance(S',S)) then
16:              S <- S'
17:              k <- 1
18:         else
19:              k <- k + 1
20:         end if
21:     end while
22:  end while
```

As we explained above, our representation of the individual has a vector of paths (Fig. 1), this vector is going to be mutated in a greedy way by the function: *Greedy-Mutation*. For every position of the vector, the function generates a random number and checks if a mutation appears. If it occurs, it executes the greedy function *getBestPath*. The *getBestPath* function uses our modified version of Yen´s algorithm and checks all possible paths with the aim of finding the best path for the current position.

4 Experimental Results and Comparisons with Other Authors

In this Section we present the experiment to adjust the α parameter. At the end of this Section we present a comparison of our metaheuristic with other approaches.

4.1 *Determining the Optimal Alpha Parameter*

In this subsection we have determined the optimal value for the α parameter. This experiment was carried out using the NSF network topology [8] and NTT network topology [9]. For further information about these test instances, please, refer to [8] and [9].

For each value of this experiment, we have performed 30 independent runs; with the purpose of ensuring a statistical relevance. Each run takes 10 minutes, uses a value of 25 for the *k-shortest-path* parameter, used in our modified version of Yen´s algorithm and six levels of neighborhood (k_{max}=6).

To decide the best value for the parameter, we have used the hypervolume concept [6]. We have used as reference point to calculate the hypervolume of each Pareto front, $r_{min} = (0,0)$ and $r_{max} = (999,99)$, where, inside the parentheses, the first value refers to y_1 and the second refers to y_2.

In Table 1, we can a see a comparison between the different varieties of VNS, in a multiobjective context. We have performed this experiment setting $\alpha = 0$ in the MO-VNS, and $\alpha = \{0.2, 0.4, 0.6, 0.8, 1\}$ in MO-SVNS. We notice that the best values for the α parameter are 0 and 0.6; both obtained very similar results in all instances. We have decided that the winner must be $\alpha = 0$ because it wins in the biggest instance ($NTT c_{ij} = 10, |U| = 40$), so the winner is MO-VNS algorithm.

Table 1 Parameter Alpha (Hypervolume)

	C_{ij}	$\|U\|$	MO-VNS	MO-SVNS				
			0	0.2	0.4	0.6	0.8	1
NSF	6	10	**0.97698**	**0.97698**	**0.97698**	**0.97698**	**0.97698**	**0.97698**
		20	**0.95896**	**0.95896**	**0.95896**	**0.95896**	**0.95896**	**0.95896**
		30	**0.93493**	**0.93493**	0.93492	**0.93493**	**0.93493**	0.93492
	8	20	**0.95896**	**0.95896**	**0.95896**	**0.95896**	**0.95896**	**0.95896**
		30	**0.93493**	**0.93493**	**0.93493**	**0.93493**	**0.93493**	**0.93493**
		40	0.89475	0.89476	**0.90387**	0.89475	0.89474	0.89477
NTT	8	10	**0.93293**	**0.93293**	**0.93293**	**0.93293**	**0.93293**	**0.93293**
		20	**0.84079**	0.83234	**0.84079**	0.84078	0.83233	0.84073
		30	0.78862	0.80511	0.7969	**0.80519**	0.78862	0.79688
	10	10	**0.93293**	**0.93293**	**0.93293**	**0.93293**	**0.93293**	**0.93293**
		20	0.84477	0.84476	0.84478	**0.84479**	0.84476	0.84477
		40	**0.70414**	0.69666	0.6958	0.6957	0.69664	0.6958
			9	7	8	9	7	6

4.2 Comparisons with Other Authors

In this subsection we compare the MO-VNS algorithm with other approaches. To be fair with the other authors, we use the same metrics and test instances as them. These metrics determinate the performance of the algorithms in quality ($M1$), distribution ($M2$) and extension of the Pareto front ($M3$), they were taken from [10]. From [5] we took the fourth metric ($M4$), it calculates the percentage of generated solutions which not belong to the Pareto front. For further information about these metrics, please, refer to [10] and [5]. Also it is presented the average of these four metrics, and it is denoted as R. In order to calculate this average the metrics $M1$ and $M4$ were slightly modified, thus, $M1^* = 1 - M1$ and $M4^* = 1 - M4$, where the best value is now 1 and the worst 0.

For each set of unicast demands U and capacities c_{ij} of wavelengths in the network, a set of optimal solutions approximated to the Pareto front are calculated using the following procedure: the algorithm was executed 10 times; a set including all the obtained solutions was generated; the dominated solutions were deleted and a new approximated set to the Pareto front was generated. It is denoted as Y_{known}.

In [11], also the number of blocked unicast request (NB) was calculated in average for the 10 runs. They presented the following algorithms [11]: BIANT (Bicriterion Ant), COMP (COMPETants), MOAQ (Multiple Objective Ant Q Algorithm), MOACS (Multi-Objective Ant Colony System), M3AS (Multiobjective Max-Min Ant System), MAS (Multiobjective Ant System), PACO (Pareto Ant Colony Optimization), MOA (Multiobjective Omicron ACO), 3SPFF (3-Shortest Path routing, First-Fit wavelength assignment), 3SPLU (3-Shortest Path routing, Least-Used wavelength assignment), 3SPMU (3-Shortest Path routing, Most-Used wavelength assignment), 3SPRR (3-Shortest Path routing, Random wavelength assignment), SPFF (Shortest Path Dijkstra routing, First-Fit wavelength assignment), SPLU (Shortest Path Dijkstra routing, Least-Used wavelength

Table 2 Comparisons with [11] using $NTT\ c_{ij} = 8$ and $|U| = \{10.20\}$

Algorithms	NTT Topology (c_{ij}=8; $\|U\|$=10)						NTT Topology (c_{ij}=8; $\|U\|$=20)					
	NB	*M1∗*	*M2*	*M3*	*M4∗*	*R10*	*NB*	*M1∗*	*M2*	*M3*	*M4∗*	*R10*
MOVNS	**0**	**1**	**1**	**1**	**1**	**1**	**0**	**0.87**	**1**	**1**	**0.083**	**0.74**
MOACS	N/A	N/A	N/A	N/A	0.05	N/A	N/A	N/A	N/A	N/A	0.07	N/A
M3AS	N/A	N/A	N/A	N/A	0	N/A	N/A	N/A	N/A	N/A	0.02	N/A
BIANT	0	0.99	0.1	0.05	0.9	0.51	0	0.82	0.95	0.71	0.02	0.62
COMP	0	0.85	0.8	0.54	0.3	0.62	0	0.74	0.98	0.76	0	0.62
MOAQ	0	0.86	1	1	0.3	0.79	0	0.69	0.9	1	0	0.65
MOACS	0	0.98	0.2	0.11	0.8	0.52	0	0.78	0.81	0.8	0	0.6
M3AS	0	1	0	0	1	0.5	0	0.82	0.86	0.77	0	0.61
MAS	0	0.98	0.2	0.1	0.7	0.49	0	0.79	1	0.86	0	0.66
PACO	0	0.99	0.1	0.05	0.9	0.51	0	0.77	0.8	0.77	0	0.59
MOA	0	1	0	0	1	0.5	0	0.82	0.94	0.88	0	0.66
3SPFF	0	0	0	0	0	0	0	0	0.03	0.06	0	0.02
3SPLU	0	1	0	0	1	0.5	0	0.35	0.08	0.07	0.65	0.29
3SPMU	0	0.03	0	0	0	0.01	0	0.08	0.11	0.09	0	0.07
3SPRR	0	1	0	0	1	0.5	0	0.36	0.14	0.1	0.1	0.17
SPFF	0	0	0	0	0	0	2	N/A	N/A	N/A	N/A	N/A
SPLU	0	1	0	0	1	0.5	2	N/A	N/A	N/A	N/A	N/A
SPMU	0	0.03	0	0	0	0.01	2	N/A	N/A	N/A	N/A	N/A
SPRR	0	1	0	0	1	0.5	2	N/A	N/A	N/A	N/A	N/A

Table 3 Comparisons with [11] using NTT $c_{ij} = 10$ and $|U| = \{20.40\}$

Algorithms	NTT Topology (c_{ij}=8; $\|U\|$=20						NTT Topology (c_{ij}=8; $\|U\|$=40					
	NB	*M1∗*	*M2*	*M3*	*M4∗*	*R10*	*NB*	*M1∗*	*M2*	*M3*	*M4∗*	*R10*
MOVNS	**0**	**0.95**	**1**	**1**	**0.35**	**0.82**	**0**	**0.90**	**1**	**1**	**0.2**	**0.77**
MOACS	N/A	N/A	N/A	N/A	0.05	N/A	N/A	N/A	N/A	N/A	0.04	N/A
M3AS	N/A	N/A	N/A	N/A	0.05	N/A	N/A	N/A	N/A	N/A	0.08	N/A
BIANT	0	0.76	0.98	0.62	0	0.59	0	0.64	1	1	0	0.66
COMP	0	0.7	0.87	1	0	0.64	0	0.46	0.82	0.91	0	0.55
MOAQ	0	0.57	0.73	0.89	0	0.55	0	0.37	0.7	0.92	0	0.5
MOACS	0	0.76	0.91	0.85	0	0.63	0	0.61	0.93	0.89	0	0.61
M3AS	0	0.75	0.8	0.6	0	0.54	0	0.52	0.88	0.88	0	0.57
MAS	0	0.6	0.81	0.84	0.02	0.57	0	0.6	0.93	0.97	0	0.62
PACO	0	0.77	0.95	0.82	0	0.63	0	0.55	0.82	0.84	0	0.55
MOA	0	0.79	1	0.84	0	0.66	0	0.6	0.87	0.88	0	0.59
3SPFF	0	0	0.1	0.15	0	0.06	0	0.3	0.41	0.3	0	0.25
3SPLU	0	0.43	0.07	0.17	0.2	0.22	0	0.86	0.38	0.26	0.13	0.41
3SPMU	0	0.05	0.13	0.15	0	0.08	0	0	0.32	0.24	0	0.14
3SPRR	0	0.39	0.12	0.2	0	0.18	0	0.81	0.31	0.24	0.08	0.36
SPFF	2	N/A	N/A	N/A	N/A	N/A	4	N/A	N/A	N/A	N/A	N/A
SPLU	2	N/A	N/A	N/A	N/A	N/A	4	N/A	N/A	N/A	N/A	N/A
SPMU	2	N/A	N/A	N/A	N/A	N/A	4	N/A	N/A	N/A	N/A	N/A
SPRR	2	N/A	N/A	N/A	N/A	N/A	4	N/A	N/A	N/A	N/A	N/A

assignment), SPMU (Shortest Path Dijkstra routing, Most-Used wavelength assignment), SPRR (Shortest Path Dijkstra routing, Random wavelength assignment).

Considering the results shown in Table 2 and Table 3, we can conclude that MO-VNS overcomes all algorithms in all instances. It seems that the use of an algorithm based on trajectory is a very promising option to solve the RWA problem.

In Table 4 we present a comparison with [12] and [13] (using M4* in %). Two ACO (Ant Colony Optimization) algorithms are presented in [12] to solve RWA problem, MOACS and M3AS. In [13], the authors have implemented a team that combines 7 different MOEAs. If we analyze the results presented in Table 4, we conclude that the MO-VNS has obtained higher results than any other.

Table 4 Comparisons with other Approaches (Metric $M4$, Average Results in %)

	MO-VNS	MOACS [12]	M3AS [12]	TA-MOEA [13]
NSF $c_{ij} = \{6,8\}$ and $\|U\| = \{(10,20,30),(20,30,40)\}$	53.88	17.33	26.33	35.5
NTT $c_{ij} = \{8,10\}$ and $\|U\| = \{(10,20,30),(10,20,40)\}$	45.55	4.68	4.93	6.1

5 Conclusions and Future Work

In this work we have developed a multiobjective trajectory-based algorithm, the Multiobjective Variable Neighborhood Search (MO-VNS) to solve the RWA problem in WDM optical networks. We also have implemented the variant Skewed Variable Neighborhood Search (MO-SVNS). After carrying out an experiment to compare both algorithms (MO-VNS and MO-SVNS), we conclude that MO-VNS obtains better results. Also, we have presented a comparison with different approaches taken from the literature. In light of the comparisons, we have realized that our metaheuristic has improved all the results obtained by more than 15 different algorithms. As future work, we have the intention of developing the algorithms NSGA-II and SPEA2, which are a standard in multiobjective context. We will also compare their results, with the ones accomplished by MO-VNS.

Acknowledgements. This work has been partially funded by the Spanish Ministry of Education and Science and ERDF (the European Regional Development Fund), under contract TIN2008-06491-C04-04 (the M* project). Álvaro Rubio-Largo is supported by the research grant PRE09010 from Junta de Extremadura (Spain).

References

1. Hamad, A.M., Kamal, A.E.: A survey of multicasting protocols for broadcast-and-select single-hop networks. IEEE Network 16, 36–48 (2002)
2. Gagnaire, M., Koubaa, M., Puech, N.: Network dimensioning under scheduled and random lightpath demands in all-optical wdm networks. IEEE Journal on Selected Areas in Communications 25(S-9), 58–67 (2007)
3. Saha, M., Sengupta, I.: A genetic algorithm based approach for static virtual topology design in optical networks. In: IEEE Indicom 2005 Conference, pp. 392–395 (2005)
4. Hansen, P., Mladenovic, N.: Variable neighborhood search: Principles and applications. European Journal of Operational Research 130, 449–467 (2001)
5. Deb, K.: Multi-Objective Optimization Using Evolutionary Algorithms. John Wiley & Sons, Inc., New York (2001)
6. Zitzler, E., Thiele, L.: Multiobjective optimization using evolutionary algorithms - a comparative case study, pp. 292–301. Springer, Heidelberg (1998)
7. Yen, J.Y.: Finding the k shortest loopless paths in a network. Manage Sci 17(11), 712–716 (2003)
8. Pinto, D., Barán, B.: Solving multiobjective multicast routing problem with a new ant colony optimization approach. In: LANC 2005: Proceedings of the 3rd international IFIP/ACM Latin American conference on Networking, pp. 11–19. ACM, New York (2005)
9. Schaerer, M., Barán, B.: A multiobjective ant colony system for vehicle routing problem with time windows. In: IASTED International Conference on Applied Informatics, pp. 97–102 (2003)
10. Zitzler, E., Deb, K., Thiele, L.: Comparison of multiobjective evolutionary algorithms: Empirical results. Evolutionary Computation 8, 173–195 (2000)
11. Arteta, A., Barán, B., Pinto, D.: Routing and wavelength assignment over wdm optical networks: a comparison between moacos and classical approaches. In: LANC 2007: Proceedings of the 4th international IFIP/ACM Latin American conference on Networking, pp. 53–63. ACM, New York (2007)
12. Insfrán, C., Pinto, D., Barán, B.: Diseño de topologìas virtuales en redes Ópticas. un enfoque basado en colonia de hormigas. In: XXXII Latin-American Conference on Informatics 2006 - CLEI 2006, vol. 8, pp. 173–195 (2006)
13. Fernandez, J.M., Vila, P., Calle, E., Marzo, J.L.: Design of virtual topologies using the elitist team of multiobjective evolutionary algorithms. In: Proceedings of International Symposium on Performance Evaluation of Computer and Telecommunication Systems - SPECTS 2007, pp. 266–271 (2007)

iGenda: An Event Scheduler for Common Users and Centralised Systems

Ângelo Costa, Juan L. Laredo, Paulo Novais, Juan M. Corchado, and José Neves

Abstract. The world is walking towards an aged society as a consequence of the increasing rate of longevity in modern cultures. With age comes the fact that memory decreases its efficiency and memory loss starts to surge. Within this context, iGenda is a Personal Memory Assistant (PMA) designed to run on a personal computer or mobile device that tries to help final-users in keeping track of their daily activities. In addition, iGenda has included a Centralised Management System (CMS) on the side of an hospital-like institution, the CMS stands a level above the PMA and the goal is to manage the medical staff (e.g. physicians and nurses) daily work schedule taking into account the patients and resources, communicating directly with the PMA of the patient. This paper presents the platform concept, the overall architecture of the system and the key features on the different agents and components.

1 Introduction

The current tendency on the growing of the population size is to decline in addition to a decreasing on the mortality rate [1]. This means that in the future there will be more elderly than young people (children and teenagers). This fact leads to the increasing of people with memory loss. Our memory is what make us who we are and help us do our everyday tasks, helping us remember the tiniest details of each

Ângelo Costa · Paulo Novais · José Neves
CCTC, Departamento de Informática, Universidade do Minho, Braga, Portugal
e-mail: acosta@di.uminho.pt, pjon@di.uminho.pt, jneves@di.uminho.pt

Juan L. Laredo
Department of Architecture and Computer Technology, University of Granada, Spain
e-mail: juanlu@geneura.ugr.es

Juan M. Corchado
Departamento Informática y Automática, University of Salamanca, Salamanca, Spain
e-mail: corchado@usal.es

E. Corchado et al. (Eds.): SOCO 2010, AISC 73, pp. 55–62.

task and keeps track of what we have done and what needs to be done. It is also a known fact that when people retire their days are filled with free time. This free time can be more harmful than beneficent. The filling of playful activities in the everyday routine can contribute to a more active ageing thus creating more interesting experiences at an old stage of life.

To cope with some of these issues (e.g. free-time management, task management,...), we have created a platform based on a Multi-Agent System that recurs on modular and collaborative agents that work on solving the Memory related problems. On top of such a platform a Personal Memory Assistant and a Social Enabler were developed and are here presented, as well as a Centralised System Manager that can manage several services that schedule several user's agenda taking into account some restrictions such as the availability of resources or the health conditions of the user. We are also going to present a validation of our choice that a distributed agent system approach is adequate for developing multi-agent systems, focusing on the PMA paradigm.

2 Ageing Factors

Nowadays because the increased quality of the health care services and the advances in the social care system the life expectancy tends to increase. In fact the United Nations (UN) stated in their last survey that "By 2050, the world as a whole and every continent except Africa are projected to have more elderly people (at least 60 years of age) than children (below 15)", in 2050 there will be twice the elderly people than nowadays [1]. There are several consequent problems inherent to the ageing of population, being the loss of memory one of the most common and often one of the most underrated. At the age of 50 the human beings begin to be affected by it, being the forgetfulness of more recent events, one of the most occurred symptoms [2]. The severity can be variable but if a person does nothing to stop it can progresses to a case of dementia. The memories lost are not recoverable and the lost capacity of memorizing is irreversible but any further loss can be prevented. Through the exercising of the brain the prevention can be done.

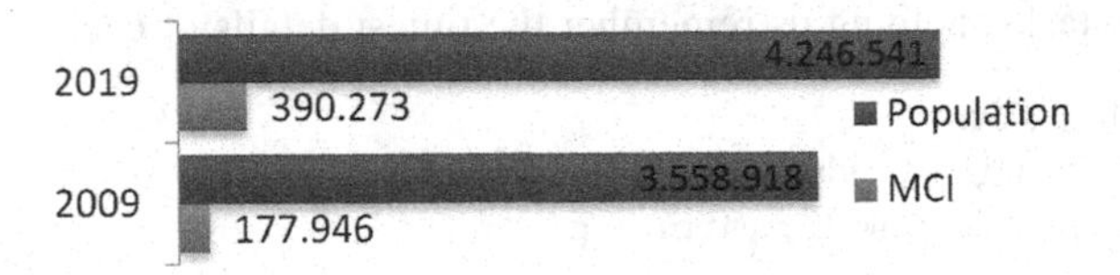

Fig. 1 Portuguese population Mild Cognitive Impairment incidence and expectancy

In fact not only the elderly people get affected by memory loss, people in a younger age range are affected too. There's not still a known way of reversing the loss of information by the human brain, so a possible solution may be the use of computational systems to store and retrieve all that data. Through the use of an agenda and/or calendar, we may reach the goals set to this work. The current technologies fail in this point, by underachieve or misinterpret the actual needs of the users or the directions to be taken in order to approach the problem [3, 4].

3 iGenda

The main objective in this work is to provide an intelligent scheduler that organizes all the user day, a Personal Memory Assistant (PMA), and a Centralised System Manager (CMS) that manages the agenda of an institution and his attendant users, and managing also the available resources. The system is in fact a two modular parts that are meant to be connected seamlessly through agents and be supported by a Multi-Agent System [5].

The PMA aims to help persons with memory loss, by sustaining all the daily events and warning the user when it is time to put them into action. It will be able to receive information delivered by any source and organise it in the most convenient way, according to predefined standards and protocols, so that the user will not need to manually plan or schedule specific events and tasks [6, 7, 8].

The CMS is aimed to manage the agenda of activities of e.g. a hospital. The management will be done taking into account the resources such as beds or offices. Every day, the CMS updates the physicist agendas with the daily planning of visits and consults. The objective is to optimize the resources and minimize the time in order to provide a better attendance to the users. The CMS is compatible with the iGenda system, it means that it is totally modular and can adapt to new events or the changing of the already appointed and if can recalculate the next consults, if the physician marks all the beginning and end of the consults.

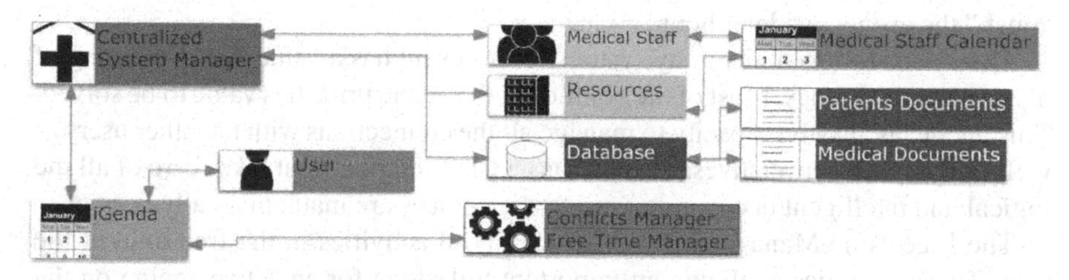

Fig. 2 Block Structure representation of the project

In addition, we present an intelligent scheduler that will help the user to remember relevant information and events, via interacting with the user through computational means, a PMA. The aim is to help specially people with memory loss, by sustaining all the daily events and remember the user to execute them.

With the capacity of receiving information delivered by any source, it proves to be a fairly modular and adaptable to any operating conditions. It takes the events received and organises them in the most convenient way so the user has to interact the least possible, having the hard work for iGenda.

The overall architecture of the system is composed by the following set of agents: the Agenda Manager (AM), the Free Time Manager (FTM), the Conflicts Manager (CM) and the Interface Manager.

The communication between agents is established by Ethernet, WIFI, GSM, UMTS, among others. The communication protocol complies with the FIPA- ACL (Agent Communication Language), being the platform JADE used to build the iGenda and the CMS architecture and using the XMLCodec in the content expressions [9]. All the agents are compliant with this standard since they are all JADE implemented agents. The agents developed in JADE-Leap [10] comply with the needed availability on mobile devices. There is also a plan of implementation of agents in JADEX platform, though in still a concept. The messages must be encrypted and secured, because agents will probably run in a low security device.

The Agenda Manager connects and controls the remaining parts of the manager system and the scheduling one, using the communication infrastructure available to receive and send requests. As result the AM stands for the communication and security of the whole systems. The AM is a two way application agent, it manages the incoming events to be scheduled and programs the time that triggers the FTM. It is also the communication relay with the rest of the system.

The AM ensures also that the user's friends and relatives can keep in touch with him and know what activities he is or will be doing.

The Conflicts Manager agent is intended to assure that there is no overlap of activities and events. This module schedules or reorganizes the events that are relayed from the Agenda Manager, insuring that they are in accordance with the rest of the events. When a collision of different events is detected, the outcome will be decided by methods of intelligent conflict management. In case of overlapping events with the same priority level, the notification of overlapping is reported to the sender, so he/she may try to reschedule to a different time slot. In case of delays the CM can "push" the events to a later hour.

The events follow a hierarchy system. Every event has a value that is defining of his priority or urgency. Most of the conflicts will use the priorities value to be solved. This agent has also the capacity to manage all the connections with the other users as well as with the user relatives. The CM has a CLP "engine" that takes care of all the logical and intelligent decisions, assuring the choices are mathematically correct.

The Free Time Manager will schedule playful activities in the free time of the user. These activities configure an important milestone for an active ageing on the part of the user, once it promotes cultural, educational and conviviality conducts, based on an individualized plan. The FTM has a database that contains information of the user's favourite activities, previously verified by the decision support group or a medical committee. The FTM connects to an activities filled database so he can retrieve the available events, relegating all the decisions to a logic engine.

The FTM uses a distribution function that decides the activity that is inserted into the user's free time. It is also important to keep in mind that the activities are merely suggestive, it comes to the user to decide to execute them or not.

The interface intends to be intuitive and easy to use. Large buttons are used and only the necessary information is displayed. When an event is triggered or

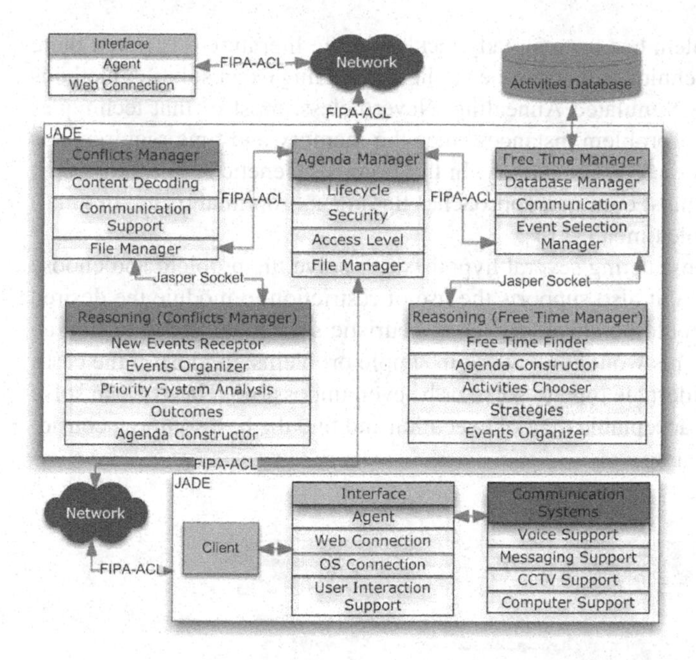

Fig. 3 Scheme of the Structure

accomplished, the user is informed. This agent assures that the information reaches the user, by computer or mobile phone, so the user is always connected to the system. The information warnings and content can be textual, in audio format (pre-recorded messages) or both. In terms of functionalities it supports the communication between the iGenda and supports also the distribution of the information through the already mentioned devices [11].

4 Centralised Management System

The CMS is designed to be capable of the integral management of an institution agenda in a Centralised way. In its current status of development, the CMS is a specification of the following requirements.

Firstly, the system has to be able to manage several agendas such as the physicists and medical staff. In addition, it has to count on the allocation of the resources, such as beds available, monitoring and operative devices. Finally, such results have to be synchronized with the final-user agendas.

Therefore, in a larger picture the iGenda will have to manage several agendas minding the fact that several resources have to be allocated to each user and they have to be available in the right moment. Hence, such a problem can be classified as a resource-constrained scheduling problem in non-stationary environments in which CMS has to react to changes in real-time.

This type of problem has been already tackled in the literature ([12]) and there are compounds of techniques that can be applied, including exact solution methods or heuristics such as Simulated Annealing. Nevertheless, most of that techniques are infeasible in large problem instances since the memory and time requirements scale in non-linear orders of complexity. In this context, Genetic Algorithms (GA) use to be one of the most common approaches finding sub-optimal (if not optimal) solutions in reasonable time [13, 14].

The GA allow constructing several hypothesis to solve the problem and choose the best case scenario, it also supports the use of restriction to module the desired goal. The problem could be solved by using heuristic search algorithm to find an optimal solution but this would only work in simple problems, in most of the cases it would take a considerable time to solve or be even impossible. The GA can solve these problems is an acceptable time of execution and find the best solution, considering all the conditionings.

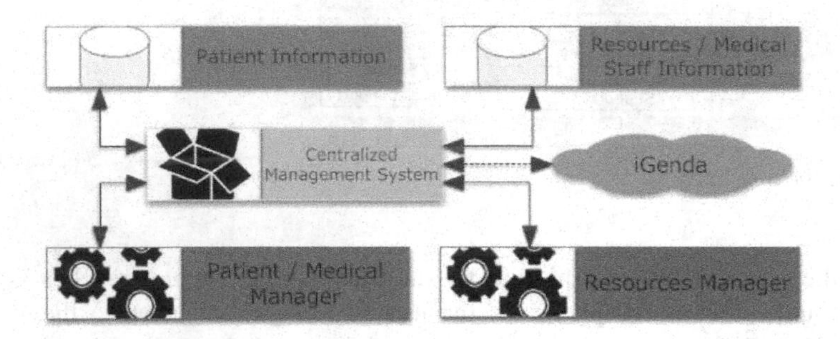

Fig. 4 CMS structure

The problem addressed will considerate feasible and infeasible requirements. The feasible are the soft requirements and the infeasible are the hard requirements. The hard requirements are:

- The health condition of the user (e.g. a heart condition).
- The available room (e.g. if the ECG room is available on a period of time).
- The available physicist (e.g. if the desired doctor is available on that day).
- The basic medical devices (e.g. the movable medical devices are available on time).

The soft requirements are:

- The available nurses.
- The preferred time (e.g. the time the user prefers to have the appointment).
- The preferred room (e.g. if the correct room is not available but a similar is).
- The time distribution (e.g. if the schedule is not optimized but serves best the most important items).

Although the main objective is to optimize the agenda in terms of minimizing the time loss and to attend the most possible patients in the least time, an important fact is not reduce the quality of service and take the desirable time with the patients, so the approximated time of attendance has to be considered to keep the standards in terms of quality of service. The operation is executed as follows:

- Initially it is gathered all the activities that must be executed in that week.
- The creation of a calendar with all that activities with little or no special scheduling.
- Detection of the most important events and marking the values of the events to prepare the next step.
- Saving the initial calendar and with the rules and values of the GA create a new calendar.
- Randomly selects parents from current population and produces new chromosomes by performing crossover operation on pair of parents.
- Randomly selects the chromosomes from current population and replaces them with new ones. The algorithm not selects chromosomes for replacement if it is among the best chromosomes in population.
- Repeat until achieved a minimal index of satisfaction.

This operation on a computer with current specifications, is aimed to run in under a second, so speed is not a problem.

The connection with the iGenda is of a crucial form. The scheduled events are automatically relayed to the patient and physicist agenda. Having the capacity of real-time additions and modifications can be beneficial and of an optimal value to the way the route the physician has to do, thus increasing the performance.

5 Conclusions and Future Work

In this paper was presented the iGenda, an event scheduler based in a multi-agent architecture. Globally the project is a joined PMA and a CMS. The objective is to organize the user's day and an institution resources and human labour. This project makes the difference to other PMAs, once it introduces the component of free time occupation. Currently most of the project is already working and the focus goes to the remaining functionalities. There have been already some tests in a closed environment, the results are under the expected spectrum, it is intended to broad the tests to a real environment.

There are still problems and critical decisions to be made, likewise the choices of technologies and algorithms must be considered because a bad choice can set-back the project several months, mostly because the re-coding.

The Centralised Management System is currently under development and the finish should be done in a few months.

In terms of future work we will consider some ideas that came up and surfaced during this work, namely:

- A Case Based Reasoning model is in study so that the iGenda will have the capacity to remember and learn from past decisions [15].
- A the weather detection mechanism will be also focus of a major overhaul in order to provide iGenda with the possibility to optimize the selection of events, considering the current and future weather (outside events).
- A Geographic Information System, a supportive GPS system that can accurately give the position of the user.
- The integration of the Centralised Management System in a real scenario, so conclusive tests can be made.

References

1. UNFPA, State of World Population 2009 (2009)
2. Tucker, G.: Age-associated memory loss: Prevalence and implications. Journal Watch Psychiatry (1995)
3. Aguilar, J.M., Cantos, J., Expósito, G., Gómez, P.: The improvement of the quality of life for elderly and relatives through two tele-assistance services: the telecare approach. In: TELECARE, pp. 73–85 (2004)
4. Brown, S.J.: Next generation telecare and its role in primary and community care. Health & Social Care in the Community 11, 459–462 (2003), (PMID: 14629575)
5. Moreno, A., Valls, A., Viejo, A.: Using jade-leap to implement agents in mobile devices (2001), http://jade.tilab.com/papers/EXP/02Moreno.pdf
6. Costa, Â., Novais, P., Costa, R., Neves, J.: Memory assistant in everyday living. In: ESM 2009 - The 23rd annual European Simulation and Modelling Conference, pp. 269–276 (2009)
7. Costa, Â., Novais, P., Costa, R., Corchado, J.M., Neves, J.: Multi-agent personal memory assistant. In: 8th International Conference on Practical Applications of Agents and Multi-Agent Systems - PAAMS 2010 (to appear, 2010)
8. Fraile, J.A., Tapia, D.I., Rodríguez, S., Corchado, J.M.: Agents in home care: A case study. In: Corchado, E., Wu, X., Oja, E., Herrero, Á., Baruque, B. (eds.) HAIS 2009. LNCS, vol. 5572, pp. 1–8. Springer, Heidelberg (2009)
9. Pokahr, A., Braubach, L.: From a Research to an Industry-Strength Agent Platform: JADEX V2 (2009)
10. Castolo, O., Ferrada, F., Camarinha-Matos, L.: Telecare time bank: A virtual community for elderly care supported by mobile agents. The Journal on Information Technology in Healthcare, 119–133 (2004)
11. Corchado, J.M., Bajo, J., Abraham, A.: Gerami: Improving healthcare delivery in geriatric residences. IEEE Intelligent Systems 23(2), 19–25 (2008)
12. Decker, K., Li, J.: Coordinated hospital patient scheduling. In: Proceedings of the Third International Conference on Multi-Agent Systems (ICMAS 1998), pp. 104–111 (1998)
13. Brucker, P., Drexl, A., Möhring, R., Neumann, K., Pesch, E.: Resource-constrained project scheduling: Notation, classification, models, and methods. European Journal of Operational Research 112(1), 3–41 (1999)
14. Laredo, J.L.J., Fernandes, C., Guervós, J.J.M., Gagné, C.: Improving genetic algorithms performance via deterministic population shrinkage. In: GECCO, pp. 819–826 (2009)
15. Luck, M., Ashri, R., d'Inverno, M.: Agent-Based Software Development. Artech House Publishers (2004)

Scalable Intelligence and Adaptation in Scheduling DSS

Ana Almeida, Constantino Martins, and Luiz Faria

Abstract. The aim of this paper is to present an evolution of a Scheduling Decision Support System (SDSS) incorporating the concepts of Scalable Intelligence and Adaptation. The purpose of our work is to contribute for DSS enhancement in the field of production plan and control, and via its practical exploitation pursuit better performance of production managers and consequent increase industrial systems efficiency and productivity.

Keywords: Adaptation, Scalable Intelligence, Decision Support System.

1 Introduction

Scheduling problems demand for information and tools that facilitate making well-informed decisions, assist in evaluating and analyzing their direct results and indirect implications, and provide strong basis for subsequent decisions. Thus Decision Support Systems (DSS) should be viewed as active participants in the decision making process. Some authors state that in areas where DSS were applied the corresponding results improved significantly. Yet, some studies point out to some constraints in Decision Support Systems (DSS) for Scheduling [1]. The first is the difficulty in an intelligent interaction between the user and the DSS. Current DSS´s are not able to explain their reasoning or proposed solutions, like expert systems, since this is very difficult for most of the scheduling algorithms. The second limitation is that DSS should not restrict to follow the user-interaction paradigm, they should infer what users are trying to do and learn from observation. Finally, decision-making is usually a process involving many people, advising for different aspects of the problem, which points for the need of collaborative approaches to scheduling process.

Ana Almeida · Constantino Martins · Luiz Faria
GECAD - Knowledge Engineering and Decision Support Group,
ISEP – Instituto Superior de Engenharia do Porto,
R. Dr. António Bernardino de Almeida, 431. 4200-072 Porto, Portugal
e-mail: {ana,const,lff}@dei.isep.ipp.pt

E. Corchado et al. (Eds.): SOCO 2010, AISC 73, pp. 63–70.

The key element of a generic adaptive system is the user model [10, 12, 13, 14]. Traditional machine learning techniques and Bayesian methods used to create user models are not enough flexible to capture the inherent uncertainty of human behavior [11]. In this context, soft computing techniques can be used to handle and process human uncertainty and to simulate human decision-making.

The aim of this paper is to present an evolution of a Scheduling Decision Support System (SDSS) incorporating the concepts of Scalable Intelligence [2] and Adaptation [3]. The former concept states that a DSS is a tool scalable in intelligence, and the user will use as much system intelligence as higher is the capacity of the system to dialogue with him adapting to his reasoning, explaining conveniently the decision making process. The later claims that adaptation/personalization can improve user transactions quality and efficiency, enhancing users contribution to the scheduling process. The practical advantages are evidenced in better performance of managers responsible for production planning and control and the consequently increased efficiency and productivity of industrial systems.

This paper is organized as follows. Section II and III presents the evolution of a Scheduling Decision Support System (SDSS) incorporating the concepts of Scalable Intelligence and Adaptation. Finally section IV presents some conclusions.

2 Scalable Intelligence

Usually, intelligent systems with Human Computer Interaction (HCI) are not Mixed-Initiative. We can find two opposite approaches [4]: user controlled interaction, where the system through a Graphical User Interface (GUI) allows the manipulation of a set of software tools, limiting itself to respond to user commands (example: scheduling systems); system controlled interaction, where users are limited to answer a set of questions addressed by the system. Scalable Intelligence (Mixed-Initiative Interaction + Intelligence) with Adaptation is a fundamental aspect to achieve an effective HCI. The Architecture diagram of the proposed system can be observed on figure 1.

3 Scheduling and Analysis

Previously we developed a prototype for production scheduling [5] to assist the manager to perform good scheduling plans, in useful time. It is uses some scheduling algorithms developed in earlier work [6], and supports user in decision making, in useful time, was improved in order to respond better to situations that was difficult to find a solution to the problem, through a Scalable Analysis. The main purpose is to endeavour to find possible schedules that fulfil due dates. When this is not possible, the Scalable Analysis suggests some overlapping alternatives, which minimize the impact over production plan accord with user-defined policy. Each suggestion comes with the respective impact analysis on the operational production plan [5].

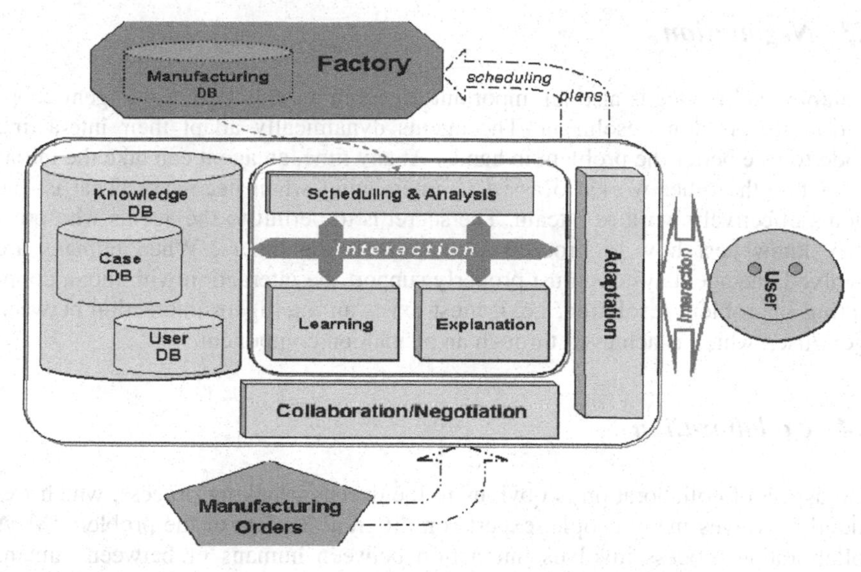

Fig. 1 Architecture diagram

3.1 Learning

In our approach to production scheduling we consider fundamental the user characteristics and his knowledge (represented in the User Model see section 3.1). We can consider three interaction modes. In the basic mode, the interaction is centred on the user due to its experience, and computer collaborates only in supporting decision-making. Afterward, as computational system acquires domain knowledge, learning from user decisions and environment variables, the interaction/decision process is balanced between them. In the advanced mode, the computer can assume the main role of the interaction/decision process, asking for user collaboration only when needs it, and also can to explain the reasoning process. This is our idea of Scalable Intelligence, that is the system being proactive it can vary the intelligence level applied to decisions according to user experience, knowledge and will.

3.2 Explanation

As an important contribution within this work we explore the explanation generation component for schedules fails. When a problem arises in schedule a manufacturing order, the system is able to explain what it is happening, based on restrictions that originated the problem and made impossible a successful scheduling satisfying all manufacturing order requisites. Consequently, the system can use this information to suggest alternatives for problem resolution, or even to do the necessary changes (rescheduling) in order to obtain a satisfactory solution for the user. This glimpses the potential of Scalable Intelligence to the success of scheduling function.

3.3 Negotiation

Scalable Intelligence is also an important aspect in an effective multi-agent cooperation for problem resolution. The agents dynamically adapt their interacting mode to face better the problem in hands. At any time, an agent can take the initiative while the other works to assist it, intervening when necessary. That is, the agents effectively work as a team. The secret is to permit to the agents who presently know best how to proceed to coordinate the others. When humans are involved, the agent system must properly support the interaction with them, cooperating in problem resolution, i.e. it must exists an intelligent interaction between agent/user, which is achieved through an adaptation component.

3.4 Collaboration

The aspect of collaboration is obvious in the decision making process, which frequently involves many people, experts on different aspects of the problem [8]. A collaboration process involves interaction between humans or between humans and machines involved in a common task, in order to share knowledge to achieve common goals.

When humans are involved in collaborative process, they are individual and social issues which have to be taken into account such as cognitive and conceptual skill, abilities, practical expertises and factual knowledge. The adaptation ability in this kind of tools, based on users profile is a key feature to provide real usefulness and effectiveness of collaborative approach especially in the scheduling context.

4 Adaptation

Adaptation focuses on users profile, providing him with appropriate information presented in the most suitable manner. This can limit information requirements and exchanging, and also the number of duration of interactions with the framework, which leads to a significant gain of the whole process.

To support effective cooperation between System and user there is the Adaptation block. Here there is an intuitive user-interface, designed to allow the user to assess the Scheduling System and to manipulate schedules down to the smallest detail. It also monitors user behaviour and adapts its presentation accordingly. User behaviour is mostly defined upon its interaction with the system itself. In our case, the Adaptation block tries to adapt the interface of the Scheduling System to the skills of the scheduler expert, reorganizing the sequence of the content presentation according to the interaction he provides. Adaptation is achieved through an User Model (UM) based on a Stereotype Model, which describes the information, knowledge and preferences of the user. There is also a Domain Model that represents concept hierarchies or maps and the related structure for the representation of the user objective and knowledge level. An Interaction Model represents and defines the adaptation between the user and the application.

4.1 User Model

Two different types of techniques can be used to implement the User Model: Knowledge and Behavioral based [9, 13]. The Knowledge-Based adaptation typically results for data collected through questionnaires and studies of the user, with the purpose to produce a set of initial heuristics. The Behavioral adaptation results from monitoring of the user during his activity [10].

Also, different methods can be used for constructing the User Model like for example [10, 13]:

- Bayesian methods;
- Machine learning methods;
- Overlay methods;
- Stereotype methods;
- Other general methods (plan recognition);
- Etc.

Bayesian networks can be use to infer user goals and decide on which actions to be taken in assistance to the user.

The use of stereotypes allows to classify users in groups and generalizes user characteristics to that group [9, 10]. The definition of the necessary characteristics for the classification in stereotypes must take to consideration the granularity degree wished [10].

The Behavioral adaptation can be implemented in two forms: the Overlay and the Perturbation methods [10]. These methods relate the level of the user knowledge with the learning objectives/competences that the user is supposed/intended to reach [10].

In the overlay method, user's knowledge is regarded as a subset of expert's knowledge. The user knowledge is a subset of the domain knowledge of the system [10].

The techniques of Overlay and stereotype can be combined [10]. The user profile is initially categorized by one stereotype but is gradually modified when the Overlay Model receives information on the interaction with the system [10].

The approach used to build ours User Model (UM) is the Stereotype Model with the overlay model for the knowledge representation of the user.

The representation of the stereotype is hierarchical. Stereotypes for user with different knowledge have been used to adapt information, interface, scenario, goals and plans.

The user modeling process starts with the identification of the user subgroup (using for example questionnaires), then the identification of key characteristics (which one to identify the members of a user-subgroup), and finally the representation in hierarchical ordered stereotypes with inheritance.

The user plan is a sequence of user actions that allows him to achieve a certain goal. The System observes the user actions ant try to infer all possible user plans. This goal is possible because our system has a library of all possible user actions and the preconditions of those actions.

A large number of criteria can be established in the Stereotype definition depending on the adaptation goals [9].

The definition of the characteristics of the User will took into account the Domain Model. For example, table 1 presents a generic profile which includes the name, age, knowledge, academics background, deficiencies and the domain of the application.

Table 1 Characteristic used in the UM

Model	Profile	Characteristics
Domain Independent Data	Generic Profile	Personal information
		Demographic data
		Academics background
		Qualifications
		Knowledge (background knowledge)
		Deficiencies: visual or others
		The Application Domain
	Psychological profile	Cognitive capacities
		Traces of the personality
		Inheritance of characteristics
Domain Dependent Data		Objectives
		Planning / Plan
		Complete description of the navigation
		knowledge Acquired
		A context model
		Aptitude
		Interests

The tools used to collect data are:

1. For the Domain Independent Data: Questionnaires, certificates and C.V., questionnaires and Psychological exams;
2. For the Domain Dependent Data: Questionnaires and exams;

4.2 Domain Model

The Domain Model represents concept hierarchies and the related structure for the representation of the user knowledge level (quantitative value).

The Domain Model use the user characteristics from the User Model (UM). The knowledge about the user, represented in the User Model, is used by the

Adaptation Model to define a specific domain concept graph, adapted from the Domain Model, in order to address the current user needs.

The path used in the graph is defined by:

1. The interaction with the user;
2. The user knowledge representation defined by the UM;
3. The user characteristics in the UM.

The results of Domain Model achieve are:

1. The development of the concept graph by each user to use in the Adaptation rules;
2. The Definition of the Adaptation Model using the characteristics of the user in the User Model.

4.3 Interaction Model

The Interaction Model represents and defines the interaction between the user and the application [9, 12, 14].

In the Interaction Model, the system presents the functionalities to change the content presentation, the structure of the links or the links annotation with the objective to allow the user to reach the goals proposed in their training. To guide the user to the relevant information and keep him away from the irrelevant information or pages that he still would not be able to understand, it is used the technique generally known by link adaptation (Hiding, disabling, removal, etc.) [9, 12]. Also, the platform supplies, in the content (page), additional or alternative information to certify that the most relevant information is shown. The technique that is used for this task it is generally known by content adaptation [9, 12].

5 Conclusions

The purpose of our work is to contribute for DSS enhancement in the field of production plan and control, and via its practical exploitation pursuit better performance of production managers and consequent increase industrial systems efficiency and productivity.

Problem resolution via Scalable Intelligence with Adaptation is perfectly tailored to human/computer integration. The future of the DSS passes for establishing intelligent interaction between users/computers self-improving its intelligence, converting themselves into truly intelligent systems.

Comparing with conventional DSS, the advantages are: Reconfigurability; Capability to explain reasoning and solutions; Better user adaptation; Learning from environment/situation; Learning from users.

These characteristics contribute to the new generation of flexible and evolutionary DSS. That is, suggest solutions to current problem and learn with the choices, adjustments and justifications from user and from obtained results. In addition, gradually refine its methods and expand its capacity and knowledge, being able to respond better to the future problems. The integration of different concepts and paradigms for the development of Scheduling DSS represents an important alternative to support the scheduling process on manufacturing environments.

References

1. Wiers, V.C.S.: Human–Computer Interaction in Production Scheduling: Analysis and Design of Decision Support Systems for Production Scheduling Tasks. Ph.D. Thesis, Eindhoven University of Technology, Holland (1997)
2. Ramos, C.: Scalable Intelligence Decision Support Systems. In: Proc. of Conf. on Enterprise Information Systems (1999)
3. Almeida, A., Martins, C.: A New Approach for Decision Support in Cooperative Scheduling. In: Kommers, P., Richards, G. (eds.) Proceedings of World Conference on Educational Multimedia, Hypermedia and Telecommunications, Orlando FL, USA, June 2006, pp. 2656–2659. AACE, Chesapeake (2006)
4. Hearst, M.: Trends & Controversies: Mixed-initiative interaction. IEEE Intelligent Systems (1999)
5. Marinho, J., Ramos, C., Oliveira, J.: SCALINTEL: A DSS Prototype with Scalable Intelligence. In: Proceedings of ICKEDS - 1st International Conference on Knowledge Engineering and Decision Support, Porto, Portugal, pp. 353–359 (2004)
6. Almeida, A.: Analysis and Development of Mechanisms and Algorithms to Support Product Oriented Scheduling. Doctoral Thesis on Production Engineering, University of Minho, Portugal (2002)
7. Silva, E., Ramos, C.: Scalable Analysis to Decision Support for Production Scheduling. In: IASTED International Conference on Artificial Intelligence and Applications (2002)
8. Almeida, A., Marreiros, G.: An approach to Collaborative Scheduling through Group Decision Support. Journal of Advanced Computational Intelligence and Intelligent Informatics 10(4) (2006)
9. Martins, C., Faria, L., Carrapatoso, E.: Constructivist Approach for an Educational Adaptive Hypermedia Tool. In: The 8th IEEE International Conference on Advanced Learning Technologies (ICALT 2008), University of Cantabria, Santander, July 1-5,(2008)
10. Martins, C., Faria, L., Carrapatoso, E.: User Modeling In Adaptive Hypermedia Educational Systems. In: Educational Technology & Society, pp. 1436–4522 (2007)
11. Frias-Martinez, E., Magoulas, G.D., Chen, S.Y., Macredie, R.D.: Modeling Human Behavior in User-Adaptive Systems: Recent Advances Using Soft Computing Technique. Expert Systems with Applications 29(2) (2005)
12. Brusilovsky, P.: From adaptive hypermedia to the adaptive Web. In: Proceedings of Mensch & Computer 2003, Stuttgart, Germany, pp. 21–24 (2003)
13. Kobsa, A.: User Modeling: Recent Work, Prospects and Hazards. In: Adaptive User Interfaces: Principles and Practice, North Holland/Elsevier, Amsterdam (1993)
14. Benyon, D.: Adaptive systems: A solution to Usability Problems. Journal of User Modeling and User Adapted Interaction 3(1), 1–22 (1993)

A Parallel Cooperative Evolutionary Strategy for Solving the Reporting Cells Problem

Álvaro Rubio-Largo, David L. González-Álvarez, Miguel A. Vega-Rodríguez, Sónia M. Almeida-Luz, Juan A. Gómez-Pulido, and Juan M. Sánchez-Pérez

Abstract. The Location Management of a mobile network is a major problem nowadays. One of the most popular strategies used to solve this problem is the Reporting Cells. To configure a mobile network is necessary to indicate what cells of the network are going to operate as Reporting Cells (RC). The choice of these cells is not trivial because they affect directly to the cost of the mobile network. Hereby we present a parallel cooperative strategy of evolutionary algorithms to solve the RC problem. This method tries to solve the Location Management, placing optimally the RC in a mobile network, minimizing its cost. Due to the large amount of solutions that we can find, this problem is suitable for being solved with evolutionary strategies. Our work consists in the implementation of some evolutionary algorithms that obtain very good results working in a parallel way on a cluster.

1 Introduction

The use of mobile networks is constantly growing. Due to the communications networks must support more and more users and their respective applications must provide a good response without losing quality and availability, it is necessary to consider the Location Management of the users when we decide to design the network infrastructure. One of the most popular strategies to solve the Location Management problem is the Reporting Cells (RC) schema proposed in [1]. In this schema a subset of cells denominated Reporting Cells exists. Each mobile terminal only performs a location update when it changes its location and moves to one of these cells. If an incoming call must be routed to a mobile user, the search must be restricted to its last reporting cell and their respective neighbour non-reporting cells. This search

Álvaro Rubio-Largo · David L. González-Álvarez · Miguel A. Vega-Rodríguez · Juan A. Gómez-Pulido · Juan M. Sánchez-Pérez
University of Extremadura, Polytechnic School, Cáceres, Spain
e-mail: {arl,dlga,mavega,jangomez,sanperez}@unex.es

Sónia M. Almeida-Luz
School of Technology and Management, Polytechnic Institute of Leiria, Portugal
e-mail: sluz@estg.ipleiria.pt

E. Corchado et al. (Eds.): SOCO 2010, AISC 73, pp. 71–78.

has an associated cost that is necessary to minimize, in this way the RC problem becomes into an NP-complete problem as it is indicated in [2]. For this reason, it is very common the use of evolutionary algorithms to solve this problem.

When we approach the resolution of a problem using evolutionary algorithms, firstly we have to select the most appropriate algorithm. It should be remembered that not all evolutionary algorithms work correctly in all kinds of problems, so if we select a not suitable algorithm for the problem we are attending, it is very likely that it does not return very good results. In this paper, we choose the following evolutionary algorithms: Genetic Algorithm (GA) [3], Greedy Randomized Adaptive Search Procedure (GRASP) [4], Differential Evolution (DE) [5], Population-Based Incremental Learning (PBIL) [6], Artificial Bee Colony (ABC) [7] and Scatter Search (SS) [8]. Recent trends in evolutionary techniques are focused on the use of a parallel cooperative strategy of evolutionary algorithms, using each skill of each algorithm [9]. The bigger the difference of all evolutionary algorithms, the better the richness will have the parallel strategy (that is, more ways for evolving). In fact, we mix population-based and trajectory-based algorithms, classical and novel algorithms, etc. In this work, we design a parallel cooperative strategy of evolutionary algorithms on a cluster computing environment to solve the RC problem, trying to combine each skill of each algorithm in order to get better results. To manage the communications among algorithms, we use the message passing interface (MPI).

This paper is organized as follows: In Section 2 we present the RC problem. In Section 3 we explain the hyperheuristic used. After some preliminary considerations (Section 4), we present all experiments and results achieved by the parallel cooperative strategy on a cluster in Section 5. Finally, we show all conclusions and future work in Section 6.

2 Reporting Cell Planning

In this section we explain the Reporting Cells scheme and how is it applied in the calculation of the location management cost.

The Reporting Cells planning was proposed by Bar-Noy and Kessler [1] with the objective of minimizing the cost of tracking mobile users. According to the Reporting Cells scheme, there are two types of cells: reporting cells (RC) and non-reporting cells (nRC), as shown in Fig. 1.

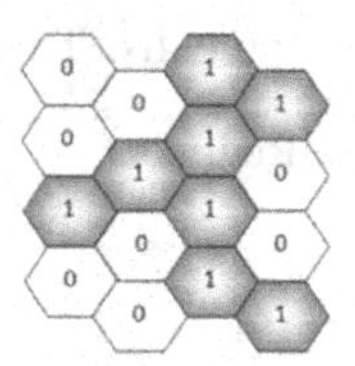

Fig. 1 RC Cells (1) and nRC Cells (0)

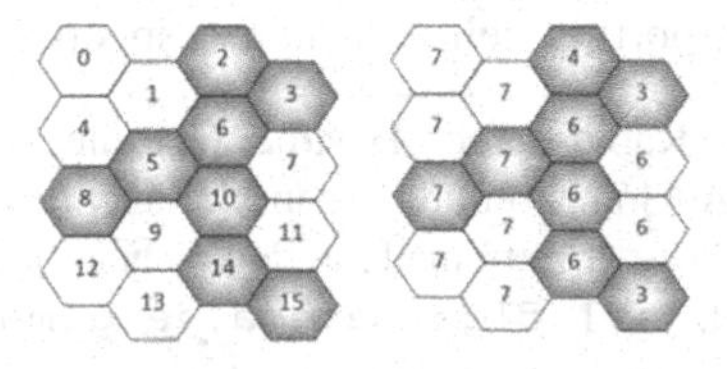

Fig. 2 Reporting Cells Planning

For all cells, RC and nRC, the vicinity factor is related by computing the maximum number of neighbour cells that must be paged if an incoming call occurs. Concretely, the vicinity factor of a single RC is given by the number of nRC accessible from this RC, without crossing any other RC and taking into account itself. On the other hand, the vicinity factor of an nRC is given by the maximum value of the vicinity factors of the RC's accessible from this nRC. This means, if an nRC has more than one neighbouring RC, it must perform this process for each of them, using the highest value of vicinity factor. As an example, in Fig. 2, the neighbouring cells of the cell 10 are the cells 7, 9, 11, 12, 13 and the own cell 10. Then, the value of the vicinity factor is 6, as it has six neighbouring cells (counting itself). On the other hand, if we see the cell 4, we have the cells 2, 5, 6 and 8 as neighbouring reporting cells, with vicinity factors 4, 7, 6, and 7, respectively. To calculate the Location Management cost we must select the highest value among them. In this case, the vicinity factor of the cell 4 is 7.

Location Management includes two elementary operations when calculating the total cost: location updates (LU) and location inquiries/paging (P), causing updating and paging costs respectively. The updating cost is caused by the sum of the costs of updating the terminals location in the network when they change their location and must register another one. The paging cost is caused by the network during a location inquiry when the network tries to locate a user and, normally, the number of paging transactions is directly related to the number of incoming calls. By using [10], we get the following generic formula to calculate the location management cost of a particular reporting cells configuration:

$$Cost = \beta * N_{LU} + N_P \tag{1}$$

The cost of the location updates is given by N_{LU}, the cost of the location paging is given by N_P, and finally, β is a constant that denotes the cost ratio of location update to a paging transaction in the network. It is proved that the updating cost is usually much higher than paging cost. For this, the updating cost is usually considered about 10 times greater than the paging cost, therefore $\beta = 10$. The generic formula above (1) must be replaced by this more detailed formula:

$$Cost = \beta * \sum_{i \in S} N_{LU}(i) + \sum_{i=0}^{N} N_P(i) * V(i) \tag{2}$$

where $N_{LU}(i)$ is the number of location updates for Reporting Cell number i, S is the set of cells defined as reporting cells, the number of arrived calls for cell i is $N_P(i)$, N is the total number of cells in the network, $V(i)$ is the vicinity factor for cell i and, finally, β is a constant representing the cost ratio of a location update to a paging transaction in the network, as described earlier. In conclusion, given a mobile network, the goal is to select which of their cells must act as RC to obtain the lowest possible cost.

3 Parallel Hyperheuristic

As we have just seen, we have developed six evolutionary algorithms that are going to be part of a working team focused on solving the RC problem. On the one hand, in this team some algorithms evolve quickly and on the other hand, other algorithms evolve more slowly but in a constant way. We are also mixing population-based and trajectory-based algorithms, classical and novel algorithms, etc. This is the main idea for developing a parallel cooperative strategy: the use of the best features of each evolutionary algorithm.

Algorithm 1. Master Pseudocode

```
1:  function Master ()
2:  Initialize the Probability Vector
3:  Generate randomly the Initial Population
4:    while (not time-limit)
5:      Assign to each core an evolutionary algorithm
6:      Spawn all processes
7:      Send the Population to all processes
8:      Receive the best result from each process
9:      Sort all results received (by quality)
10:     Update the Probability Vector
11:   end while
12: end
```

We use a cluster with 16 nodes and a total of 128 cores (8 cores per node). We designed a parallel hyperheuristic that optimize its resources. As it is shown in the pseudocode, the initial idea consists in distributing all evolutionary algorithms among the cores, occupying all of them except one, which will be used by the Master process. This process is in charge of controlling and synchronizing all slave processes. First of all, we distribute our six evolutionary algorithms in a fair way for all slave processes, to do this, we have a probability vector that indicates us which is the probability that an evolutionary algorithm will be assigned to a core. Initially, all the algorithms have the same probability. Then, the master process distributes each evolutionary algorithm to each core, according to the probability vector. To do that, it generates a random number (between 1-100) and, depending on this number, the appropriate algorithm will be assigned to the first core, we repeat this for all cores. Therefore, using this method it is probable that the initial distribution will not be totally homogeneous. To ensure a minimal involvement of each evolutionary algorithm in each iteration, we establish that there must be at least five cores assigned to each algorithm, so in the delivery section only 97 cores are going to participate. Once assigned all evolutionary algorithms to their corresponding cores, they start executing, they will be working until the alarm rings (time limit is expired), when it happens, they must synchronize with the master process, sending to it their best result. When the master process gets all "best results" from all slave processes, it sorts them by quality to update the probability vector, recalculating all probabilities for every evolutionary algorithm. The result set gathered by the master process

(127 solutions/individuals) will be the initial population for all the algorithms when they will be restarted. Depending on the kind of algorithm (trajectory or population-based), one or more results will be sent.

The algorithm that provides better results to the parallel cooperative strategy will have more probabilities to be assigned in more cores in the cluster. To do this, we have to take into account only the 50 best results (out of 127, selective pressure) provided by all the slave processes in each synchronization. The way to update the probability vector is to add a 2% per each occurrence of an algorithm in these 50 best results. For example, if an algorithm has 18 occurrences in the 50 best results, its probability will be 36%.

4 Considerations and Test Networks Used

In this section we present: the best configurations obtained for each evolutionary algorithm of the parallel strategy, the technical characteristics of the cluster used and the different test networks.

After an exhaustive series of experiments with each algorithm, the best configuration is obtained with the following parameters:

Algorithm	Parameter	Value
GA [3]	Crossover Probability	0.5
	Tournament selection	2 individuals
	Probability (of the best individual) in the tournament	0.6
	Number of Elitist Samples	1
	Mutation Probability	0.01
	Maximal Number of Cycles with Stagnation	20
	Huge Mutation Probability	0.5
GRASP [4]	Initialization Probability	0.75
	Size of the Restricted Candidate List	$\sqrt{NumberofCells}$
	Constructor	RG Constructor
DE [5]	Crossover Factor	0.15
	Mutation Factor	0.5
	DE scheme	DE/Rand/1/bin
PBIL [6]	Number of Vectors to Update from	2
	Learning Rate	0.1
	Mutation Probability	0.075
	Mutation Shift	0.075
	Extension	Equitative MSolutions
ABC [7]	Mutation Factor	0.05
	Local Search for the Scout Bees	5bits
	Number of Scout Bees	1
SS [8]	Reference Set Size	10
	B1 (best samples)	9
	B2 (dispersed samples)	1

With these configurations we obtain, for every algorithm, very similar results to those described in [2], [10], and [11], using the same test networks. The population size in all the algorithms is set to 127, except for the GRASP that is a trajectory-based algorithm.

The cluster used in the experiments has the following specifications: 16 nodes with 8 cores (Intel Xeon 2.33 GHz) and 8 GB of DDR2 RAM (in total: 128 cores and 128GB of DDR2 RAM), all nodes have installed Scientific Linux 5.3 (64 bits) and MPICH2 v1.0.8. The test networks used to adjust each evolutionary algorithm of the parallel strategy have been obtained from [2], [10], and [11]. The sizes of these networks are: 16, 36, 64 and 100 cells. Later, we developed a larger network [12] to exploit the parallel strategy to the fullest, using a network of 900 cells.

5 Experiments and Comparisons

In this section we show the results of the experiments and the performance comparison between each different evolutionary algorithm and the parallel strategy. For each experiment, and each combination of parameters, we perform 30 independent executions of 30 minutes to ensure the statistical relevance of the results. In all these experiments we use a network of 900 cells [12].

The synchronization time between the master process and slave processes must be a parameter to adjust, because it may influence on the achievement of better results. We use the best configuration of each algorithm (see Section 4) to obtain the optimal synchronization time. This experiment was run using the following values (elapsed time between synchronizations): 2 minutes, 5 minutes, 10 minutes and 15 minutes; because the total run time of the parallel strategy was 30 minutes. These results can be seen in Table 1. With this experiment we conclude that if we synchronize the processes every 5 minutes, the parallel strategy evolves toward better solutions. This is because with this synchronization value, each evolutionary algorithm has enough working time and there is an acceptable number of synchronizations between the master process and slave processes.

With the parallel strategy configured (synchronizations every 5 minutes), we can make different comparisons. As we can see in Table 2, the parallel strategy performs better than any individual algorithm. However, we see how the majority of algorithms are unable to improve their solutions by themselves, needing help from others to continue evolving. Unlike evolutionary algorithms, the parallel strategy does not suffer stagnation. In Table 2 we can observe how the parallel strategy is

Table 1 Synchronization adjustment results (in cost units $*10^3$). We present the average (Avg.), the best (Best) and the standard deviation (Std.) on 30 independent execution.

		Minutes														
		2	4	6	8	10	12	14	16	18	20	22	24	26	28	30
Sync 2m	Avg.	5723.1	5556.3	5448.0	5404.2	5378.5	5367.6	5361.9	5359.5	5358.7	5358.5	5358.5	5358.4	5358.2	5357.9	5357.5
	Best	5674.6	5502.2	5409.0	5364.9	5317.3	5307.2	5302.4	5300.9	5300.1	5300.1	5300.1	5300.1	5300.1	5300.1	5300.1
	Std.	18.9	22.9	29.2	21.2	26.6	27.6	28.3	28.4	28.2	28.1	28.1	28.0	27.9	27.5	26.8
Sync 5m	Avg.	**5710.5**	**5361.7**	**5327.1**	**5320.8**	**5313.3**	**5305.3**	**5303.4**	**5299.1**	**5298.5**	**5298.4**	**5297.7**	**5297.7**	**5297.5**	**5297.5**	**5297.5**
	Best	5681.7	5345.7	5317.0	5308.6	5301.6	5296.9	5296.7	5294.9	5294.4	5294.3	5293.7	5293.7	5293.7	5293.7	5293.7
	Std.	14.7	7.5	5.4	4.3	5.2	5.7	4.0	4.0	3.9	2.7	2.6	2.4	2.3	2.2	2.2
Sync 10m	Avg.	5719.8	5442.5	5340.5	5328.8	5325.7	5318.8	5317.7	5311.6	5306.1	5303.5	5300.5	5299.9	5299.5	5299.2	5298.9
	Best	5672.4	5418.3	5327.3	5316.4	5315.3	5312.7	5307.4	5299.3	5297.4	5296.8	5295.3	5295.1	5294.8	5294.6	5294.5
	Std.	16.9	11.6	6.7	5.3	4.9	3.7	4.5	5.6	5.0	4.1	3.3	3.3	3.4	3.4	3.3
Sync 15m	Avg.	5723.6	5443.6	5339.7	5329.7	5327.3	5325.9	5325.1	5318.9	5318.1	5311.7	5304.7	5302.0	5300.5	5299.6	5299.3
	Best	5694.7	5416.9	5328.5	5321.6	5319.1	5318.9	5318.9	5312.1	5309.9	5301.2	5297.9	5297.3	5296.5	5296.0	5295.6
	Std.	13.8	10.7	5.1	4.5	3.9	3.5	3.2	3.3	3.4	4.7	3.8	2.9	2.5	2.3	2.1

Table 2 Empirical results (in cost units $*10^3$) for a comparison between the 6 evolutionary algorithms and our hyperheuristic (HH) every 2 minutes.

		Minutes														
		2	4	6	8	10	12	14	16	18	20	22	24	26	28	30
HH	Avg.	**5710.5**	**5361.7**	**5327.1**	**5320.8**	**5313.3**	**5305.3**	**5303.4**	**5299.1**	**5298.5**	**5298.4**	**5297.7**	**5297.7**	**5297.5**	**5297.5**	**5297.5**
	Best	5681.7	5345.7	5317.0	5308.6	5301.6	5296.9	5296.7	5294.9	5294.4	5294.3	5293.7	5293.7	5293.7	5293.7	5293.7
	Std.	14.7	7.5	5.4	4.3	5.2	5.7	4.0	4.0	3.9	2.7	2.6	2.4	2.3	2.2	2.2
GA	Avg.	6683.4	6356.1	6225.2	6153.7	6102.6	6066.0	6031.6	6006.1	5982.9	5963.2	5946.1	5930.7	5917.0	5905.0	5896.9
	Best	6405.4	6234.2	6134.9	6091.9	6045.6	5981.5	5965.4	5943.7	5929.9	5887.9	5850.1	5819.2	5775.5	5771.6	5769.8
	Std.	165.9	68.8	44.3	37.0	37.2	40.5	39.9	37.8	39.0	42.3	40.8	42.5	44.0	42.6	40.8
GRASP	Avg.	6916.4	6758.9	6709.1	6682.6	6645.2	6618.2	6595.1	6578.0	6563.0	6562.3	6552.2	6546.4	6536.5	6535.8	6525.9
	Best	6453.9	6453.9	6453.9	6453.9	6453.9	6453.9	6419.4	6419.4	6418.2	6418.2	6418.2	6418.2	6371.5	6371.5	6371.5
	Std.	300.7	176.8	151.8	129.2	124.4	113.4	114.2	113.9	118.6	118.2	94.3	84.9	88.0	87.7	81.7
DE	Avg.	5749.9	5387.4	5342.9	5338.0	5335.9	5334.7	5333.9	5333.4	5333.2	5333.0	5332.9	5332.7	5332.6	5332.5	5332.5
	Best	5707.4	5341.6	5323.1	5317.1	5315.5	5314.6	5314.6	5314.6	5314.6	5314.6	5314.6	5314.6	5314.6	5314.6	5314.6
	Std.	25.2	19.2	7.4	6.5	6.0	5.7	5.6	5.5	5.5	5.4	5.4	5.4	5.4	5.4	5.4
PBIL	Avg.	6047.6	5991.8	5979.0	5970.9	5963.8	5959.8	5956.3	5953.6	5952.6	5950.2	5947.4	5947.1	5947.0	5945.7	5944.7
	Best	5968.7	5936.1	5936.1	5936.1	5934.0	5922.8	5922.8	5922.8	5922.8	5891.8	5891.8	5891.8	5891.8	5891.8	5891.8
	Std.	41.1	24.1	22.0	15.5	14.5	17.0	15.7	16.2	15.4	18.3	18.4	18.4	18.2	18.0	17.4
ABC	Avg.	5760.6	5687.3	5647.0	5623.9	5605.3	5588.7	5575.0	5565.8	5556.0	5548.0	5540.5	5535.0	5528.7	5524.1	5521.1
	Best	5713.0	5608.9	5583.1	5565.5	5554.7	5542.3	5532.1	5529.2	5519.0	5512.4	5506.5	5501.3	5488.8	5484.4	5480.2
	Std.	27.0	24.6	24.6	24.7	24.1	22.2	19.7	17.3	16.7	14.5	15.9	15.6	16.3	15.8	16.1
SS	Avg.	6129.2	5669.6	5550.1	5497.4	5470.9	5458.5	5452.0	5448.6	5446.3	5445.4	5445.0	5444.9	5444.7	5444.5	5444.5
	Best	5970.8	5619.0	5495.7	5452.6	5430.8	5420.3	5416.5	5414.2	5414.2	5414.2	5414.2	5414.2	5414.2	5414.2	5414.2
	Std.	101.4	34.6	24.4	19.5	18.3	18.3	17.7	17.7	17.4	17.7	17.7	17.6	17.7	17.6	17.5

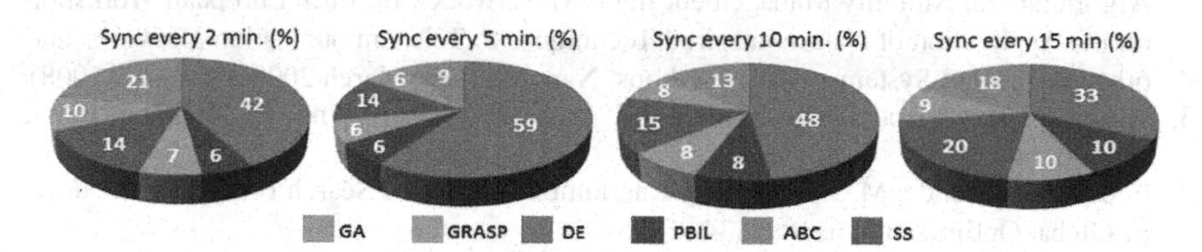

Fig. 3 Participation of each algorithm in the parallel strategy in different synchronizations

still evolving toward a better solution, although more slowly in the final stretch. These results suggest that the optimum fitness for the network of 900 cells used is close to the value reached by the parallel strategy.

Fig. 3 shows the participation rates of each algorithm at the different synchronizations. Each pie graph was generated from the distribution made by the master process after each synchronization. In the synchronizations every 2 and 15 minutes, we can see how the distribution is done more homogeneously than in the synchronizations performed every 5 and 10 minutes. In Table 1, we can see how performing the synchronizations every 5 and 10 minutes we obtain a faster evolution at the beginning of the process. Therefore, a large number of DE processes encourage the parallel strategy.

6 Conclusions and Future Work

In this work we have proposed the use of cluster computing and a parallel cooperative strategy of evolutionary algorithms to solve the RC problem. The objective of our strategy has been to try to exploit the different skills of each algorithm, we have achieved the best configuration of each of them getting very good results. Moreover, we have demonstrated that the use of a parallel strategy achieves better results than those obtained by individual algorithms. After several tests we can conclude that the best results are obtained by performing synchronizations every 5 minutes.

As future work we pretend to test the parallel strategy with new instances of our problem. We also intend to add new algorithms to our parallel strategy, trying to improve the results obtained so far. On the other hand, we will try to explore other areas of computing as Grid computing.

Acknowledgements. This work was partially funded by the Spanish Ministry of Science and Innovation and ERDF (European Regional Development Fund), under the contract TIN2008–06491–C04–04 (the M* Project). Álvaro Rubio-Largo and David L. González-Álvarez are supported by research grants from Junta de Extremadura and Fundación Valhondo, respectively.

References

1. Bar–Noy, A., Kessler, I.: Tracking mobile users in wireless a communication networks. In: INFOCOM, pp. 1232–1239 (2003)
2. Alba, E., García–Nieto, J., Taheri, J., Zomaya, A.: New Research in Nature Inspired Algorithms for Mobility Management in GSM Networks. In: Fifth European Workshop on the Application of Nature–inspired Techniques to Telecommunication Networks and other Connected Systems, EvoWorkshops, Napoles, Italy, March 2008, pp. 1–10 (2008)
3. Holland, J.H.: Adaptation in Natural and Artificial Systems. Univ. of Michigan Press, Ann Arbor (1975)
4. Feo, T.A., Resende, M.G.C.: Greedy Randomized Adaptive Search Procedures. Journal of Global Optimization 6, 109–134 (1995)
5. Price, K., Storn, R.: Differential Evolution – A Simple Evolution Strategy for Fast Optimization. Dr. Dobbs Journal 22(4), 18–24, 78 (1997)
6. Baluja, S.: Population-Based Incremental Learning: A method for integrating genetic search based function optimization and competitive learning. Technical Report CS–94–163, Carnegie Mellon University (1994)
7. Karaboga, D., Basturk, B.: A powerful and efficient algorithm for numerical function optimization: artificial bee colony (ABC) algorithm. Journal of Global Optimization 39, 459–471 (2007)
8. Glover, F.: A Template for Scatter Search and Path Relinking. In: Hao, J.-K., Lutton, E., Ronald, E., Schoenauer, M., Snyers, D. (eds.) AE 1997. LNCS, vol. 1363, pp. 1–51. Springer, Heidelberg (1998)
9. Segura, C., et al.: Optimizing the DFCN Broadcast Protocol with a Parallel Cooperative Strategy of Multi–Objective Evolutionary Algorithms. In: Ehrgott, M., Fonseca, C.M., Gandibleux, X., Hao, J.-K., Sevaux, M. (eds.) EMO 2009. LNCS, vol. 5467, pp. 305–319. Springer, Heidelberg (2009)
10. Almeida–Luz, S.M., Vega–Rodríguez, M.A., Gómez–Pulido, J.A., Sánchez–Pérez, J.M.: Applying Differential Evolution to the Reporting Cells Problem. In: International Multiconference on Computer Science and Information Technology (IMCSIT 2008), Wisla, Poland, October, pp. 65–71 (2008)
11. Subrata, R., Zomaya, A.: A Comparison of Three Artificial Life Techniques for Reporting Cell Planning in Mobile Computing. IEEE Transactions on Parallel and Distributed Systems 14(2), 142–153 (2003)
12. Rubio–Largo, A., González–Álvarez, D.L., Vega–Rodríguez, M.A.: Test Networks for RC, `http://arco.unex.es/rc`

Optimization of Parallel Manipulators Using Evolutionary Algorithms

Manuel R. Barbosa, E.J. Solteiro Pires, and António M. Lopes

Abstract. Parallel manipulators have attracted the attention of researchers from different areas such as: high-precision robotics, machine-tools, simulators and haptic devices. The choice of a particular structural configuration and its dimensioning is a central issue to the performance of these manipulators. A solution to the dimensioning problem, normally involves the definition of performance criteria as part of an optimization process. In this paper the kinematic design of a 6-dof parallel robotic manipulator for maximum dexterity is analyzed. The condition number of the inverse kinematic jacobian is defined as the measure of dexterity and solutions that minimize this criterion are found through a genetic algorithm formulation. Subsequently a neuro-genetic formulation is developed and tested. It is shown that the neuro-genetic algorithm can find close to optimal solutions for maximum dexterity, significantly reducing the computational load.

1 Introduction

Parallel manipulators are well known for their high dynamic performances and low positioning errors [1]. In the last few years parallel manipulators have attracted great attention from researchers involved with robot manipulators, robotic end effectors, robotic devices for high-precision tasks, machine-tools, simulators, and haptic devices.

One of the most important factors affecting the performance of a robotic manipulator is its structural configuration. The kinematics, statics, dynamics and

Manuel R. Barbosa · António M. Lopes
IDMEC – Pólo FEUP, Faculdade de Engenharia da Universidade do Porto,
Rua Dr. Roberto Frias 4200–465 Porto, Portugal
e-mail: {mbarbosa,aml}@fe.up.pt

E.J. Solteiro Pires
Centro de Investigação e de Tecnologias Agro-Ambientais e Biológicas,
Escola de Ciências e Tecnologia da Universidade de Trás-os-Montes e Alto Douro,
Quinta de Prados, 5000–911 Vila Real, Portugal
e-mail: epires@utad.pt

E. Corchado et al. (Eds.): SOCO 2010, AISC 73, pp. 79–86.
springerlink.com

stiffness are all dependent upon it. Following the definition of a particular structural configuration, the next step in the manipulator design is its dimensioning. Usually this task involves the choice of a set of parameters that define the mechanical structure of the manipulator. The parameter values should be chosen to optimize some performance criterion, dependent upon the foreseen application [2]. In the last years most of the research in parallel manipulators optimization has been done over several criteria related to the optimization of the manipulator workspace [3]. Other authors choose to optimize the structural stiffness of the manipulator, as this is one of the main advantages of a parallel configuration over a serial one [4]. Finally, some works may be referred where the optimization criteria are related with the manipulability, or dexterity, of the manipulator [5].

Optimization can be a difficult and time-consuming task, because of the great number of optimization parameters and the complexity of the objective functions. However, optimization procedures based on evolutionary approaches have been proved as an effective way out [6].

In this paper the kinematic design of a 6-dof parallel robotic manipulator for maximum dexterity is analyzed. First, the condition number of the inverse kinematic jacobian is used as a measure of dexterity and a Genetic Algorithm (GA) is used to solve the problem. The highly nonlinear nature of the problems involved and the normally associated time consuming optimization algorithms used, can be naturally approached by artificial Neuronal Networks (NNs) techniques. In order to explore the advantages of NNs and GAs, a neuro-genetic formulation is developed and tested. It is shown that the neuro-genetic algorithm can find close to optimal solutions for maximum dexterity, significantly reducing the computational burden. The error involved is believed to be less significant when extending the approach to a multi-objective and global optimization problem.

2 Parallel Manipulator Structure and Kinematics

The mechanical structure of the parallel robot comprises a fixed (base) platform and a moving (payload) platform, linked together by six independent, identical, open kinematic chains (Figure 1a). Each chain comprises two links: the first link (linear actuator) is always normal to the base and has a variable length, l_i, with one of its ends fixed to the base and the other one attached, by a universal joint, to the second link; the second link (arm) has a fixed length, L, and is attached to the payload platform by a spherical joint. Points B_i and P_i are the connecting points to the base and payload platforms. They are located at the vertices of two semi-regular hexagons, inscribed in circumferences of radius r_B and r_P, that are coplanar with the base and payload platforms. The separation angles between points B_1 and B_6, B_2 and B_3, and B_4 and B_5 are denoted by $2\phi_B$. In a similar way, the separation angles between points P_1 and P_2, P_3 and P_4, and P_5 and P_6 are denoted by $2\phi_P$ (Figure 1a).

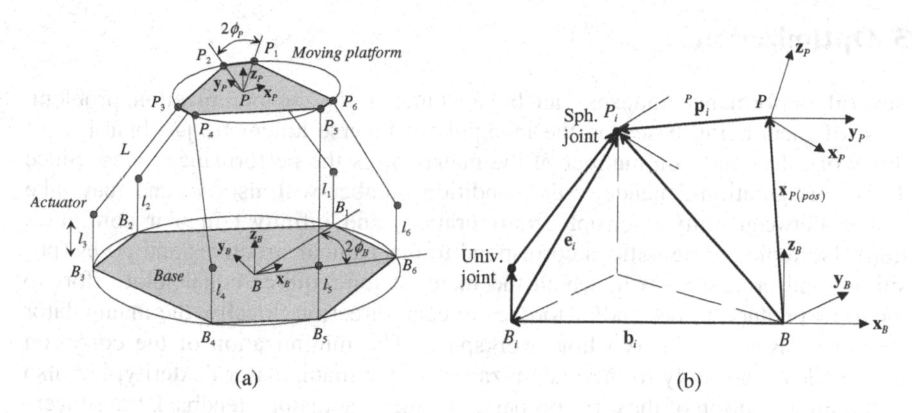

Fig. 1 Schematic representation of the parallel manipulator structure (a); Schematic representation of a kinematic chain (b)

The robot's inverse velocity kinematics can be represented by the inverse kinematic jacobian, $\mathbf{J}_C$, relating the joints velocities to the moving platform Cartesian-space velocities (linear and angular). The inverse jacobian matrix is given by equation (1) which can be computed using vector algebra [7].

$$\mathbf{J}_C = \begin{bmatrix} \dfrac{(\mathbf{e}_1 - l_1\mathbf{z}_B)^T}{(\mathbf{z}_B^T\mathbf{e}_1 - l_1)} & \dfrac{({}^P\mathbf{p}_1\,|_B \times (\mathbf{e}_1 - l_1\mathbf{z}_B)^T)}{(\mathbf{z}_B^T\mathbf{e}_1 - l_1)} \\ \vdots & \vdots \\ \dfrac{(\mathbf{e}_6 - l_6\mathbf{z}_B)^T}{(\mathbf{z}_B^T\mathbf{e}_6 - l_6)} & \dfrac{({}^P\mathbf{p}_6\,|_B \times (\mathbf{e}_6 - l_6\mathbf{z}_B)^T)}{(\mathbf{z}_B^T\mathbf{e}_6 - l_6)} \end{bmatrix} \quad (1)$$

All vectors are obtained analyzing each kinematic chain of the parallel manipulator (Figure 1b). Vectors $\mathbf{e}_i$ are given by

$$\mathbf{e}_i = \mathbf{x}_{P(pos)} - \mathbf{b}_i + {}^P\mathbf{p}_i\,|_B \quad (2)$$

where $\mathbf{x}_{P(pos)}$ is the position of the moving platform expressed in the base frame, $\mathbf{b}_i$ represent the positions of points B_i and ${}^P\mathbf{p}_i\,|_B = {}^B\mathbf{R}_P \cdot {}^P\mathbf{p}_i$ are the positions of points P_i, on the moving platform, expressed in the base frame. Matrix ${}^B\mathbf{R}_P$ represents the moving platform orientation matrix. The scalars l_i are the actuators' displacements, given by

$$l_i = e_{iz} + \sqrt{L^2 - e_{ix}^2 - e_{iy}^2} \quad (3)$$

3 Optimization

Several performance indexes can be formulated for the optimization problem, most of them being based on the manipulator inverse kinematic jacobian [2]. In this work the condition number of the matrix $\mathbf{J}_C$ is the performance index. Since $\mathbf{J}_C$ is configuration-dependent, the condition number will also be, and may take values between unity (isotropic configuration) and infinity (singular configuration). Isotropic configurations correspond to mechanical structures and poses (positions and orientations) in which the manipulator requires equal joint effort to move or produce forces and/or torques in each direction. Ideally, the manipulator should be isotropic in its whole workspace. The minimization of the condition number leads not only to the maximization of the manipulator dexterity, but also to the minimization of the error propagation due to actuators, feedback transducers and, when $\mathbf{J}_C$ matrix is inverted, numerically induced noise [8]. Mathematically, the condition number is given by

$$\kappa = \frac{\sigma_{max}(\mathbf{J}_C)}{\sigma_{min}(\mathbf{J}_C)} \tag{4}$$

where $\sigma_{max}(\mathbf{J}_C)$ and $\sigma_{min}(\mathbf{J}_C)$ represent the maximum an minimum singular values of matrix $\mathbf{J}_C$.

To obtain a dimensionally homogeneous inverse jacobian matrix, the manipulator payload platform radius, r_P, is used as a characteristic length. Thus, the same 'cost' will be associated to translational and rotational movements. The inverse kinematic jacobian results dependent upon ten variables, four of them are manipulator kinematic parameters: the position and orientation of the payload platform; the base radius (r_B); the separation angles on the payload platform (ϕ_P); the separation angles on the base (ϕ_B); and the arm length (L). We will consider a particular manipulator pose, corresponding to the centre of the manipulator workspace i.e., $[0\ 0\ 2\ 0\ 0\ 0]^T$ (units in r_P and degrees, respectively). Thus, for this pose, the inverse jacobian matrix will be a function of the four kinematic parameters.

3.1 Genetic Algorithm Based Approach

The robot kinematic parameters are codified by real values through the vector $\mathbf{p} = [r_B\ \phi_P\ \phi_B\ L]^T$. The solutions are randomly initialized in the range $1.0 \le r_B \le 2.5$, $0° \le \phi_P, \phi_B \le 25°$ and $2.0 \le L \le 3.5$. The algorithm has a tournament-2 selection to determine the parents of the offspring [9]. After selection, the simulated binary crossover and mutation operators with $p_c = 0.6$ and $p_m = 0.05$ probabilities, respectively, are called [10]. At the end of each cycle, it is used a $\varepsilon-(\lambda+\mu)$ strategy to select the solutions which survive for the next iteration. This means the best solutions among parents and offspring are chosen. At this stage, the space is divided in hyper-planes separated by the distance ε and all solutions that fall into two consecutive hyper-planes are considered having the same preference, even if their fitness values are different [11]. Two consecutive hyper-planes define a rank.

In order to sort the solutions in a rank, the maximin sorting scheme is used [12]. The ε value is initialized with 20 and is decreased during the evolution, until it reaches the value 0.03. The ε is decreased by 90% every time the best 200 solutions belong to the first rank and the ε value has not changed during the last 100 generations. The solutions are classified by the fitness function given by equation (4), in case the solution is admissible, otherwise the value 1×10^{20} is assigned. The global results (Figure 2) show multiple sets of optimal parameters. Moreover, the algorithm draws a representative solution set of the optimal parameters front.

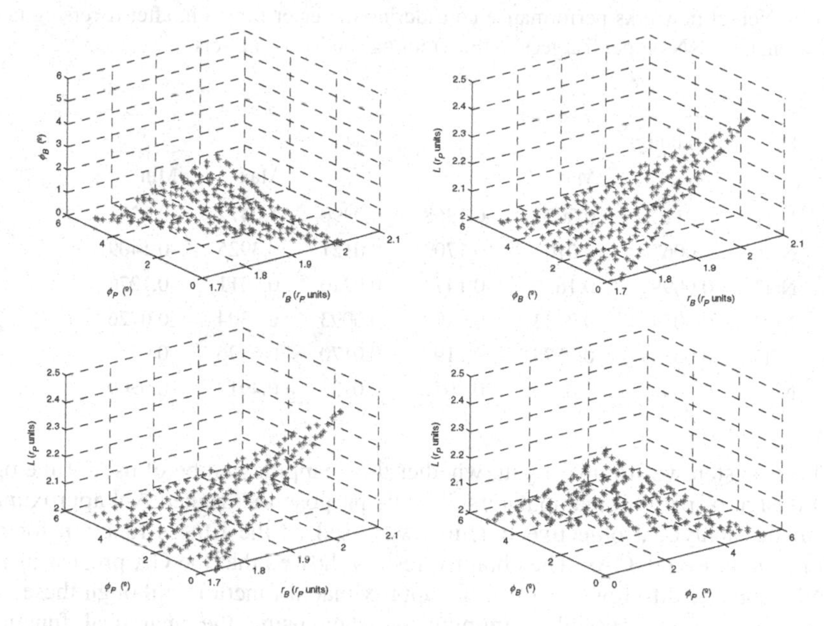

Fig. 2 Simulation results: optimal sets of kinematic parameters

3.3 Neuro-genetic Algorithm Based Approach

The process of designing a NN solution to a specific problem is mainly guided by trial and experimentation, due to the lack of explicit and proven methods that can be used to choose and set the various parameters involved in the NN design process. Among the different structures and types of NN available [13] the experiments were done using the multilayer feedforward NN architecture and using the Levenberg-Marquardt learning algorithm. The representation of the problem in this structure consisted of defining the kinematic parameters (r_B, ϕ_P, ϕ_B, L) as four input elements to the network, and κ as the output element. The design process then consisted of training and testing networks with different numbers of elements in the hidden layer (i.e., 25, 50 and 100). The data used was randomly generated and divided in three data sets (70% training, 15% validation, 15% testing). Normalization was applied to both input and output values. The criteria chosen to stop the training process was the number of successive increases of the error on the validation set (validation checks) and different values were experimented in order to visualize the performance of the network through the mean square error function (*mse*). Considering the range of values of the objective function included in the data sets [1.4170, 12.0140], the results obtained (Table 1) with mapping the objective function using NNs show that a good approximation is possible in average terms (i.e., $\sqrt{mse} < 0.01$), but extreme error values can be larger by more than one order of magnitude (i.e., 0.33), although in a few cases.

Table 1 Neural networks performance considering the error function, after reversing normalization, (i.e., NN output-Target) on the Training and Test data sets

	Training			Test		
	$\sqrt{mse}$	Max	Min	$\sqrt{mse}$	Max	Min
Net1	0.0112	0.3091	-0.1368	0.0290	1.0472	-0.1237
Net2	0.0067	0.0702	-0.1709	0.0121	0.3925	-0.1409
Net3	0.0099	0.1624	-0.1473	0.0220	0.7411	-0.1276
Net4	0.0034	0.0543	-0.0490	0.0093	0.3304	-0.0726
Net5	0.0055	0.0578	-0.1197	0.0170	0.6196	-0.0744
Net6	0.0049	0.0550	-0.1041	0.0138	0.4912	-0.0809

The next step was to investigate whether this mapping can be of use for the optimal design of parallel manipulators. To this purpose the trained NN approximation to the analytical objective function was used as the fitness function for an optimization through GAs. The obtained results showed that the GA process using the NN converged to low values of the approximated function, although these are not as close to the possible minimum as when using the analytical function (Figure 3). For the 200 best possible values obtained in each search experiment using NNs, the respective values for the approximated objective function were always well above the range [1.41, 1.42], obtained using the analytical function. The best neuro-genetic approach (GA-Net2) was based on the smaller sized network (Net2) which seems to indicate the higher importance of the generalization ability compared to the approximation error. Moreover, Figure 3 seems to confirm that NNs are good as interpolators, but not as good for extrapolation. The NNs ability to map a function, based on sets of data from that function, can not be expected to produce good maps of extreme values of the function if they have not been included in the training sets. Therefore, at most it can be expected for them to map close to minimum values.

A plot of the real values, i.e., obtained from the analytical function, and the respective NN output, is made for each of the 200 best cases (Figure 4). Although the quality of the solutions obtained are still above the overall minimum range values [1.41, 1.42], more cases are closer, even if more dispersed, to these values. In spite of the limitations revealed by the NNs based solutions, in the ability to extrapolate well to the minimum values of a function, they may be useful when a

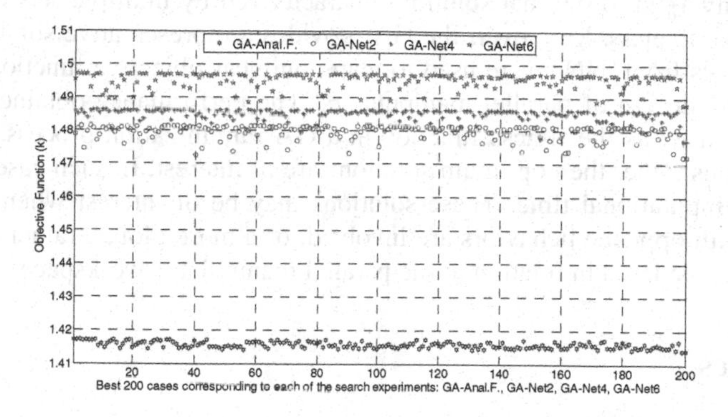

Fig. 3 Values of the objective function (κ) for each of the 200 considered best cases: GA-Analytical Func., GA-Net2, GA-Net4, GA-Net6

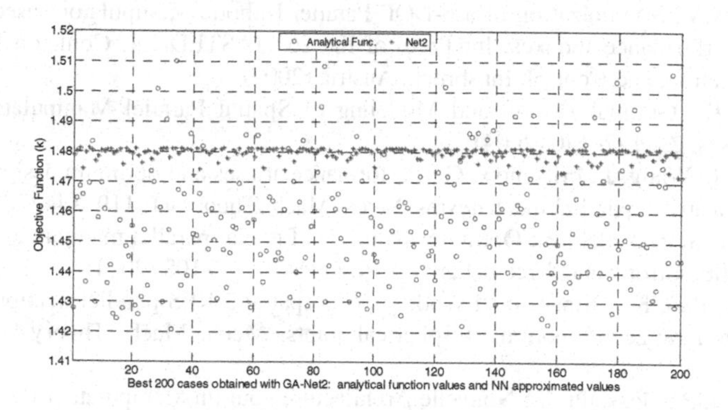

Fig. 4. Values of the objective function (κ) for each of the 200 considered best cases using GA-Net2: values of the analytical function and values of the NN approximation

multi-objective and global optimization problem is considered. In such cases, more than considering absolute minimum values for different objective functions, recognizing patterns of good solutions may provide better and more feasible approaches. The ability to provide good solutions for each objective function, as the NN based solutions provide, rather than the optimum solutions can therefore be equally important. Another issue related to the use of a GA-NN based approach was the reduction of the search time by 30% to 50% compared to the use of a GA-analytical function.

5 Conclusions

In this paper the kinematic design of a 6-dof parallel robotic manipulator for maximum dexterity was analyzed. First, a GA was used to solve the optimization problem. Afterwards, a neuro-genetic formulation was developed and tested.

The GA converged to optimal solutions characterized by multiple sets of optimal kinematic parameters. Moreover, the GA provides a representative solution set of the parameters front. NNs were used to map nonlinear objective functions associated with the design of parallel manipulators. The performance obtained showed they can be used as the fitness function in a GA minimization process, provided good solutions rather then optimum solutions are of interest. In such cases they reduce the computational time. These solutions may be of interest when objective functions with opposite behaviors are involved, or a more global, rather than local problem is considered in relation to the parallel manipulator workspace.

References

1. Chablat, D., Wenger, P., Majou, F., et al.: An Interval Analysis Based Study for the Design and the Comparison of Three-Degrees-of-Freedom Parallel Kinematic Machines. Int. J. Robot Res. 23, 615–624 (2004)
2. Lopes, A.M.: Optimization of a 6-DOF Parallel Robotic Manipulator based on Kinematic Performance Indexes. In: Proc. of the 26th IASTED Int. Conf. on Modelling, Identification, and Control, Innsbruck, Austria (2007)
3. Miller, K.: Optimal Design and Modeling of Spatial Parallel Manipulators. Int. J. Robot Res. 23, 127–140 (2004)
4. Liu, X.-J., Wang, J., Pritschow, G.: Performance atlases and optimum design of planar 5R symmetrical parallel mechanisms. Mech. Mach. Theory 41, 119–144 (2006)
5. Alici, G., Shirinzadeh, B.: Optimum synthesis of planar parallel manipulators based on kinematic isotropy and force balancing. Robotica 22, 97–108 (2004)
6. Rao, N., Rao, K.: Dimensional synthesis of a spatial 3-RPS parallel manipulator for a prescribed range of motion of spherical joints. Mech. Mach. Theory 44, 477–486 (2009)
7. Merlet, J.-P., Gosselin, C.: Nouvelle Architecture pour un Manipulateur Parallele a Six Degres de Liberte. Mech. Mach. Theory 26, 77–90 (1991)
8. Yoshikawa, T.: Manipulability of Robotic Mechanisms. Int. J. Robot Res. 4, 3–9 (1985)
9. Michalewicz, Z., Fogel, D.B.: How to solve it: modern heuristics. Springer, New York (2000)
10. Deb, K.: Multi-Objective Optimization Using Evolutionary Algorithms. John Wiley & Sons, Chichester (2001)
11. Laumanns, M., Thiele, L., Deb, K., et al.: Archiving with Guaranteed Convergence and Diversity in Multi-Objective Optimization. In: Proc. of the Genetic and Evolutionary Comp. Conference, San Francisco, USA (2002)
12. Solteiro Pires, E.J., Mendes, L., de Moura Oliveira, P.B., et al.: Single-objective front optimization: application to RF circuit design. In: Proc. of the 10th annual conference on Genetic and evolutionary computation, Atlanta, USA (2008)
13. Gupta, M., Jin, L., Homma, N.: Static and Dynamic Neural Networks, From Fundamentals to Advanced Theory. John Wiley & Sons, Chichester (2003)

Multi-criteria Manipulator Trajectory Optimization Based on Evolutionary Algorithms

E.J. Solteiro Pires, P.B. de Moura Oliveira, and J.A. Tenreiro Machado

Abstract. This paper proposes a method, based on a genetic algorithm, to generate smoth manipulator trajectories in a multi-objective perspective. The method uses terms proportional to the integral of the squared displacements in order to eliminate the jerk movement. In this work, the algorithm, based on NSGA-II and maximin sorting schemes, considers manipulators of two, three and four rotational axis (2*R*, 3*R*, 4*R*). The efficiency of the algorithm is evaluated, namely the extension of the front and the dispersion along the front. The effectiveness and capacity of the proposed approach are shown through simulations tests.

1 Introduction

Genetic algorithms (GAs) are one of the most popular evolutionary inspired search and optimization technique. This popularity is shown by the large number of successful applications in many scientific areas [1]. One of the advantages of GAs over classical techiques is that it can be adopted in optimization applications without requiring specific knowledge about the working problem. Initialy, it was mainly used in single-objective problems. However, in few years became clear that GAs could be applied in multi-objective optimization. Taking advantage of using a population, GA can determine a set of non-dominated solutions in just one execution of the algorithm [2, 3, 4, 5]. Moreover, GAs are less susceptible to the Pareto front shape or continuity than classical optimization techniques. Due to these factors,

E.J. Solteiro Pires · P.B. de Moura Oliveira
Centro de Investigação e de Tecnologias Agro-Ambientais e Biológicas,
Escola de Ciências e Tecnologia da Universidade de Trás-os-Montes e Alto Douro,
Quinta de Prados, 5000–801 Vila Real, Portugal
e-mail: {epires,oliveira}@utad.pt

J.A. Tenreiro Machado
Instituto Superior de Engenharia do Porto, Instituto Politécnico do Porto,
4200–072 Porto, Portugal
e-mail: jtm@isep.ipp.pt

E. Corchado et al. (Eds.): SOCO 2010, AISC 73, pp. 87–94.
springerlink.com

multi-objective evolutionary algorithms (MOEAs) have become increasingly popular in a vast number of areas. For example in electrical engineering, hydraulics, robotics, control scheduling, physics, medicine and computer science [6].

In robotics, the problems of trajectory planning, collision avoidance and manipulator structure design considering a single criteria has been solved using several techniques [7, 8, 9, 10]. However, trajectory planning adopting multiple objectives was somewhat overlooked and only a few articles analyzing this topic can be found. Pires *et al.* [11] proposed a MOEA to optimize a manipulator trajectory considering multiple objectives, namely: space and joint arm displacements and the energy required to perform the route without coliding with the obstacles. Ramabalan *et al.* [12] proposed two MOEAs to generate a manipulator trajectory with multiple objectives. In their work, they compare the results obtained by different algorithms. Liu *et al.* [13] considers the planning of a space manipulator taking acount the joint angle, joint velocity and torque constraints. They use a weighted fitness function where the weights are ramdomly selected. They decompose the trajectory considering several segments. In each segment a suitable polinomial is used. The inter-knots parameters (angle, velocity and torque) of each trajectory segment are optimized by the proposed MOEA.

In this line of thought, this paper proposes the use of a multi-objective method to optimize the trajectory of manipulators with 2, 3 and 4 rotational degrees of freedom (dof). The method is based on the NSGA-II and the maximin sorting scheme [14] adopting the direct kinematics. The non-dominated trajectories are those that minimize the joint and gripper displacement. In a second phase, non-dominated solutions are analyzed in order to measure the solution spread along the front.

The paper is organized as follows. Section 2 introduces the problem and the GA based scheme for its resolution. Sections 3 describes the method to measure the solution spread along the Pareto front. Based on this formulation, section 4 presents the results for several simulations involving 2, 3, and 4 link manipulators. Finally, section 5 outlines the main conclusions.

2 Problem Formulation

This work considers robotic manipulators with 2, 3 and 4 links which are required to move between two coordinates (Figure 1) in the operational space. The work considers two objectives, namely the joint displacement (O_q) and the gripper displacement (O_p). It is intended to determine a representative set of non-dominated solutions belonging to the Pareto optimal front. To measure the quality of the algorithm, the solutions spread along the front is analyzed. The decision maker chooses the solution taking into account the compromise between the objectives that he finds more appropriate.

The experiments consist on moving a iR robotic arm, $i = \{2,3,4\}$, between configurations defined by the points $A \equiv \{1.2, -0.3\}$ and $B \equiv \{-0.5, 1.4\}$. To find out the initial and final $2R$ manipulator configurations the inverse kinematics is used. For the $3R$ and $4R$ manipulators, the initial joint values are determined to obtain a

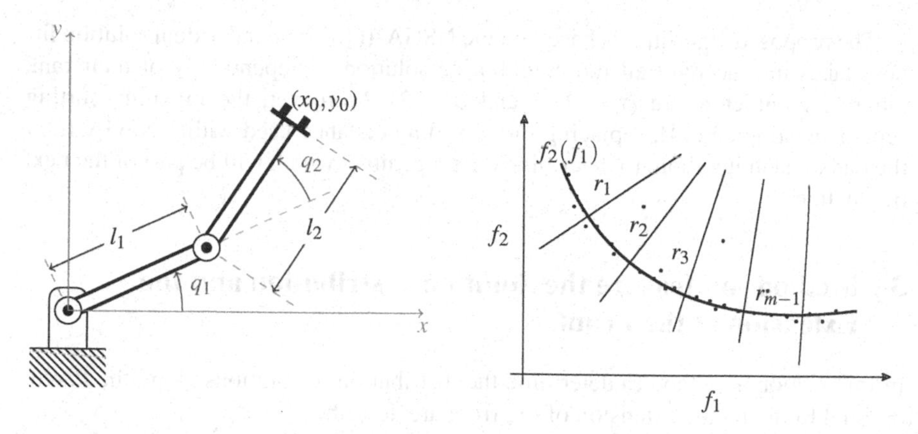

Fig. 1 Two joint (link) manipulator (2*R* robot) **Fig. 2** Straight lines normal to the front obtained

configuration close to the one observed for the 2*R* robot. In the work, the rotational joints are free to rotate 2π rad.

The GA parameters are: population size $pop_{\text{dim}} = 300$, number of generations $T_{\text{t}} = 1500$; crossover and mutation probabilities $p_{\text{c}} = 0.6$ and $p_{\text{m}} = 0.05$, respectively, link length $l = 2/i$ [m] and mass of $m = 2/i$ [Kg].

The two indices $\{O_q, O_p\}$ presented in (1) quantify the quality of the evolving trajectories of the robotic manipulators. The indices represent the joint displacement, O_q (1a), and the Cartesian gripper displacement O_p (1b):

$$O_q = \sum_{j=1}^{n} \sum_{l=1}^{i} \left(\dot{q}_l^{(j\Delta t, T)} \right)^2 \tag{1a}$$

$$O_p = \sum_{j=2}^{n} d\left(p_j, p_{j-1}\right)^2 \tag{1b}$$

The path for a *iR* manipulator ($i = 2, 3, 4$), at generation T, is directly encoded as a string in the joint space to be used by the GA. This string is represented by expression (2), where i represents the number of dof and Δt the sampling time between two consecutive configurations. Therefore, one potential solution is encoded as:

$$[\{q_1^{(\Delta t,T)}, \ldots, q_i^{(\Delta t,T)}\}, \ldots, \{q_1^{(2\Delta t,T)}, \ldots, q_i^{(2\Delta t,T)}\}, \ldots, \{q_1^{((n-2)\Delta t,T)}, \ldots, q_i^{((n-2)\Delta t,T)}\}] \tag{2}$$

where the joints values $q_l^{(j\Delta t,0)}$, $j = 1, \ldots, n-2$; $l = 1, \ldots, i$, are randomly initialized in the range $]-\pi, +\pi]$ [rad]. The robot movement is described by $n = 8$ configurations. However, the initial and final configurations are not encoded into the string, because they remain unchanged throughout the trajectory search. Without losing generality, for simplicity, it is adopted a normalized time of $\Delta t = 0.1$ s, but it is always possible to perform a time re-scaling.

The proposed algorithm is based on the NSGA-II [3]. The individual solution fitness takes into account all the neighboring solutions independently of their rank (sharing parameters are $\sigma = 0.01$ and $\alpha = 2$). Moreover, the maximin sorting scheme is adopted [14], replacing the crowding distance used within NSGA-II, at the end of each iteration, to determine the progenitors which will be part of the next population.

3 Method to Measure the Solution Distribution and the Extension of the Front

In this section a method to determine the distribution of solutions along the Pareto optimal front and the extension of the front are described.

The method begins by finding a function that models the Pareto optimal front in the appropriate range. This function should be valid between the extreme solutions obtained by the MOEA. Next, the function adopted to represent the Pareto front is divided through normal straight lines (figure 2). Between each two consecutive normal straight lines (r_i and r_{i+1}) a range I_i is defined. Finally, all solutions located in a specific range are counted. The dispertion is given by the solution number of the ranges.

The extension of the front is measured through the curve length of the modeled function taking into account the two closest points to the two extreme solutions of the Pareto front found by the MOEA.

4 Results

This section develops several tests. For each type of test multiple experiments are performed to achieve $n_{\text{exp}} = 21$ valid simulations. This means that many distinct experiments are executed until 21 successful convergences to the optimal Pareto front are obtained, for the $2R$, $3R$ and $4R$ manipulators. Figure 3 depicts one optimal Pareto front for each manipulator type. The convergence rates were 95%, 57% and 38% for the $2R$, $3R$ and $4R$ manipulators, respectively. It can be observed that as the number of links increases the convergence rates decrease. This is due to the exponential increase of the number of local fronts, with the number of robot links.

In all cases, the fronts can be modeled by equation (3) with the parameter set $\{\kappa, \alpha, \beta\} \in \mathbb{R}$. For each front obtained in a valid simulation,the parameters are estimated. Next, the median, mean and standard deviation of the parameters κ, α and β are calculated (Table 1). The front modelling procedure though equation 3 is only valid between the non-dominated extreme solutions a and b.

$$O_p(O_q) = \kappa \frac{O_q + \alpha}{O_q + \beta} \tag{3}$$

From Table 1 it can be seen that the mean and the median values are almost similar. Additionally, the values of the standard deviation are relatively small, leading to the

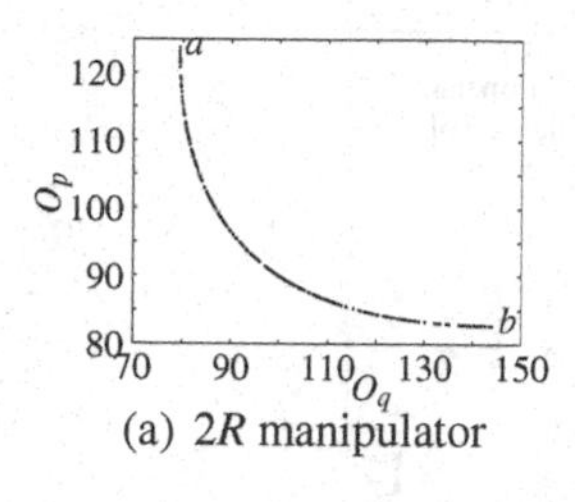

(a) 2R manipulator

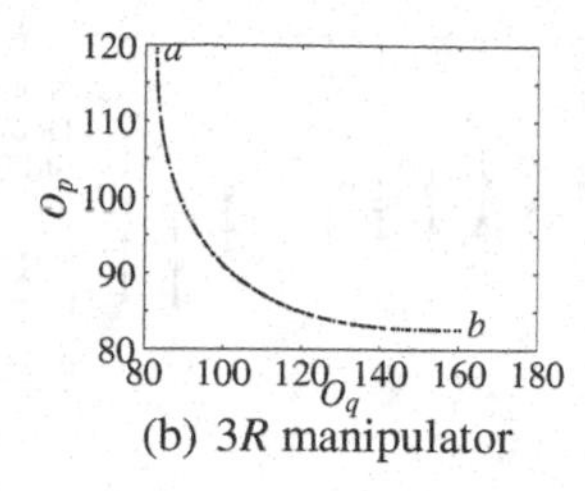

(b) 3R manipulator

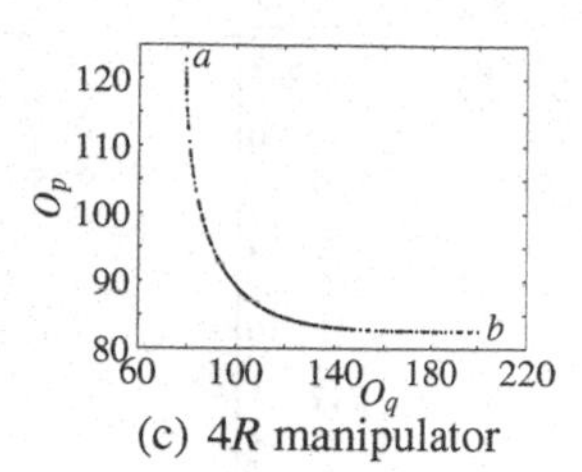

(c) 4R manipulator

Fig. 3 Pareto optimal fronts

Table 1 Statistical parameters for the modeled fronts

	2R robot front			3R robot front			4R robot front		
	κ	α	β	κ	α	β	κ	α	β
Median	77.90	−66.83	−71.21	78.68	−71.36	−75.12	80.28	−71.82	−74.44
Mean	77.99	−66.74	−71.13	78.76	−71.33	−75.19	80.15	−71.49	−74.34
Std Dev.	0.44	0.71	0.42	0.58	1.67	1.22	0.60	2.21	1.62

conclusion that the algorithm always converges to the same front, which is likely to be the Pareto optimal front. Moreover, with the increasing of the manipulator number of links the standard deviation increases. By other words, as the number of links increases the problem complexity becomes higher and the algorithm has more difficulty to converge to the same non-dominated front.

The extension (*i.e.*, the length) of the front for the iR robot manipulators, $i = \{2,3,4\}$, has an average $\mu_{\text{Ext}} = \{86.57, 105.61, 200.49\}$ and a standard deviation $\sigma_{\text{Ext}} = \{2.00, 17.48, 48.94\}$. It can be concluded that, with the increasing problem complexity, the algorithm has more difficulty in obtaining always the same front extension.

The solution diversity along the non-dominated front is presented in figures 4-6. The final percentage average of solution number belonging to the non-dominated front is $\mu_{\text{Div}} = \{96.15\%, 92.53\%, 88.25\%\}$ and the standard deviation is $\sigma_{\text{Div}} = \{1.26, 5.21, 4.28\}$. The figures reveal that the solutions are distributed over all intervals. However, this distribution is not uniform and the uniformity decreases as the number the manipulator links increases. This phenomena occurs because the solutions percentage of the non-dominated front decreases with the number of links. This is due to the fact that the algorithm favors non-dominated solutions, and only then enters into account with diversity. Consequently, the proposed algorithm reveals its potential only when all population elements are within the non-dominated front. Figures 4-6 also show that regions with less non-dominated solutions are compensated by the algorithm with more dominated solutions, in order to keep a good solution distribution in all intervals.

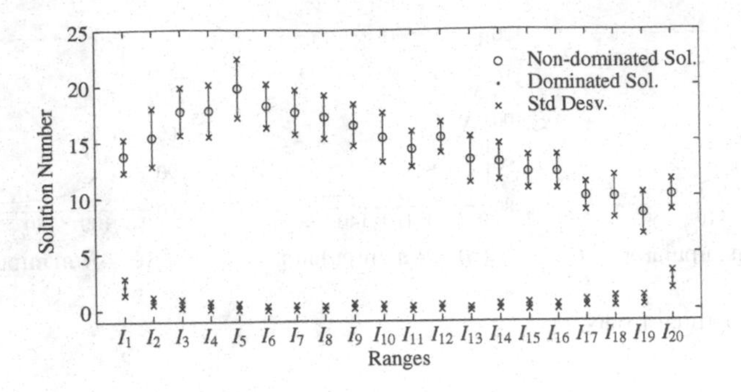

Fig. 4 The 2*R* robot solution distribution along the Pareto front

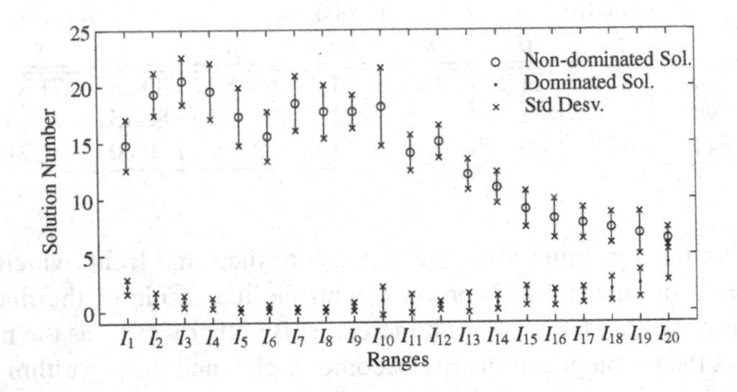

Fig. 5 The 3*R* robot solution distribution along the Pareto front

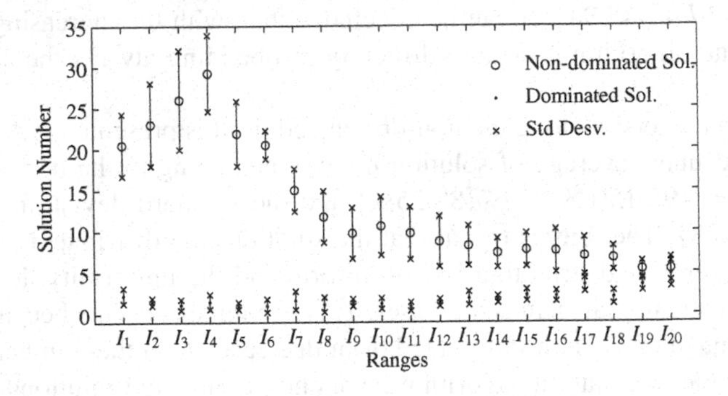

Fig. 6 The 4*R* robot solution distribution along the Pareto front

Figure 7 depicts the extreme solutions, *a* and *b*, corresponding to the optimal fronts Pareto illustrated in Figure 3. The figure includes the successive configurations and the angular displacements for the 2*R*, 3*R* and 4*R* manipulators.

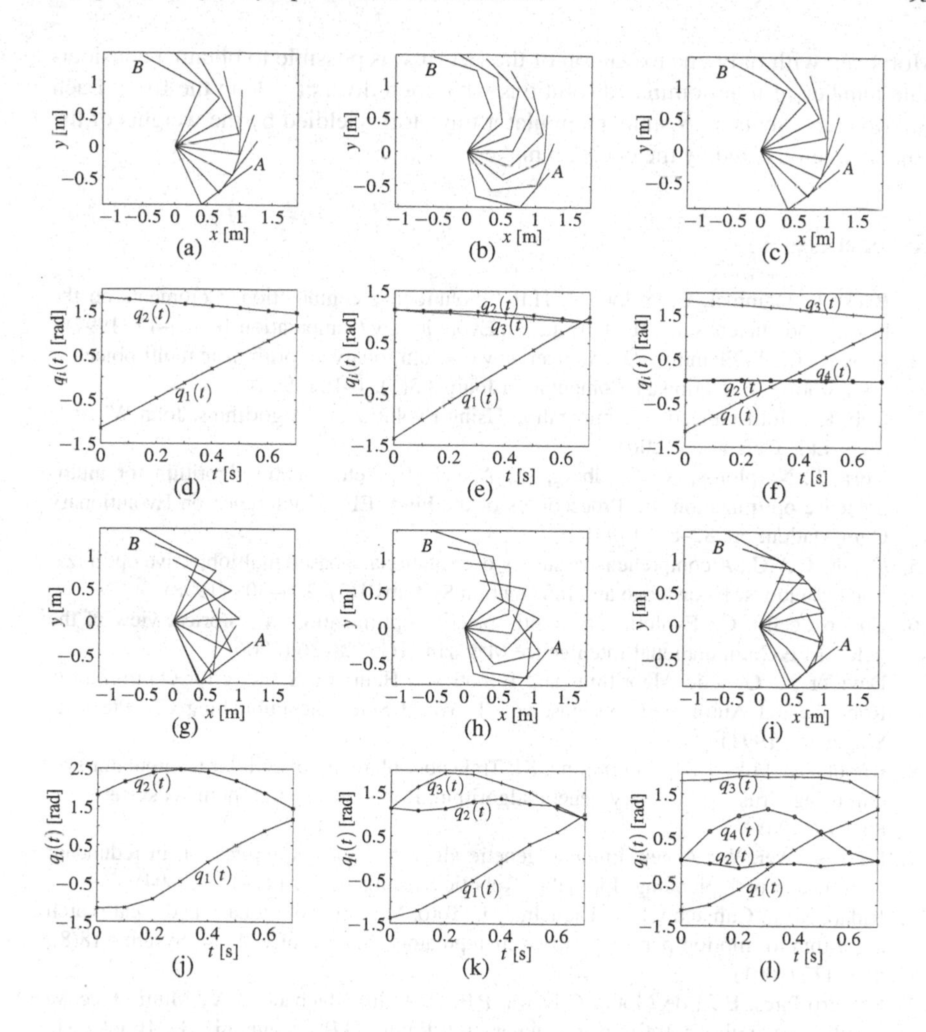

Fig. 7 Pareto optimal trajectories. Successive configurations of solution a (Figures 3(a)-3(c)) for robot: (a) $2R$, (b) $3R$, (c) $4R$. Successive configurations of solution b for robot: (g) $2R$, (h) $3R$, (i) $4R$. Joint position trajectory *vs.* time of solution a for robot: (d) $2R$, (e) $3R$, (f) $4R$. Joint position trajectory *vs.* time of solution b for robot: (j) $2R$, (k) $3R$, (l) $4R$.

5 Conclusions

This paper solves the manipulator trajectory planning problem in a multi-objective perspective. This work considered manipulators with two, three and four rotational joints with trajectories solved by an evolutionary algorithm based on NSGA-II and maximin sorting schemes. To study the efficiency of the algorithm the extension of the front and the dispersion along the front were carried out. The results show that it is possible obtain different solutions according to the weight of the objectives.

Moreover, with only one execution of the GA, it was possible to obtain a considerable number of non-dominated solutions with good diversity along the front. Each solution represents a possible manipulator trajectory yieldied by the weights of the objectives envisaged by the decision maker.

References

1. Bäck, T., Hammel, U., Schwefel, H.P.: Evolutionary computation: Comments on the history and current state. IEEE Trans. on Evolutionary Computation 1(1), 3–17 (1997)
2. Fonseca, C.M., Fleming, P.J.: An overview of evolutionary algorithms in multi-objective optimization. Evolutionary Computation Journal 3(1), 1–16 (1995)
3. Deb, K.: Multi-Objective Optimization Using Evolutionary Algorithms. John Wiley & Sons, Ltd., Chichester (2001)
4. Horn, J., Nafploitis, N., Goldberg, D.: A niched pareto genetic algorithm for multi-objective optimization. In: Proceedings of the First IEEE Conference on Evolutionary Computation, pp. 82–87 (1994)
5. Coello, C.A.C.: A comprehensive survey of evolutionary-based multiobjective optimization techniques. Knowledge and Information Systems 1(3), 269–308 (1998)
6. Coello Coello, C.: Evolutionary multi-objective optimization: a historical view of the field. IEEE Computational Intelligence Magazine 1(1), 28–36 (2006)
7. Davidor, Y.: Genetic Algorithms and Robotics, a Heuristic Strategy for Optimization. Robotics and Automated Systems, vol. 1. World Scientific Publishing Co. Pte Ltd., Singapore (1991)
8. Kubota, N., Fukuda, T., Shimojima, K.: Trajectory planning of cellular manipulator system using virus-evolutionary genetic algorithm. Robotics and Autonomous systems 19, 85–94 (1996)
9. Luo, X., Wei, W.: A new immune genetic algorithm and its application in redundant manipulator path planning. Journal of Robotic Systems 21(3), 141–151 (2004)
10. Ridao, M.A., Camacho, E.F., Riquelme, J., Toro, M.: An evolutionary and local search algorithm for motion planning of two manipulators. Journal of Robotic Systems 18(8), 463–476 (2001)
11. Solteiro Pires, E.J., de Moura Oliveira, P.B., Tenreiro Machado, J.A.: Multi-objective genetic manipulator trajectory planner. In: Raidl, G.R., Cagnoni, S., Branke, J., Corne, D.W., Drechsler, R., Jin, Y., Johnson, C.G., Machado, P., Marchiori, E., Rothlauf, F., Smith, G.D., Squillero, G. (eds.) EvoWorkshops 2004. LNCS, vol. 3005, pp. 219–229. Springer, Heidelberg (2004)
12. Ramabalan, S., Saravanan, R., Balamurugan, C.: Multi-objective dynamic optimal trajectory planning of robot manipulators in the presence of obstacles. The International Journal of Advanced Manufacturing Technology 41(5-6), 580–594 (2009)
13. Liu, Z., Huang, P., Yan, J., Liu, G.: Multi-objective genetic algorithms for trajectory optimization of space manipulator. In: 4th IEEE Conference on Industrial Electronics and Applications, ICIEA 2009, Xi'an, China, May 25-27, pp. 2810–2815 (2009)
14. Solteiro Pires, E.J., de Moura Oliveira, P.B., Tenreiro Machado, J.A.: Multi-objective MaxiMin Sorting Scheme. In: Coello Coello, C.A., Hernández Aguirre, A., Zitzler, E. (eds.) EMO 2005. LNCS, vol. 3410, pp. 165–175. Springer, Heidelberg (2005)

Combining Heuristics Backtracking and Genetic Algorithm to Solve the Container Loading Problem with Weight Distribution

Luiz Jonatã Pires de Araújo and Plácido Pinheiro

Abstract. We approach the container loading problem with maximization of the weight distribution. Our methodology consists of two phases. In the first phase, it applies heuristics based on integer linear programming to construct blocks building of small items. A backtracking algorithm chooses the best heuristics. The objective of this phase is to maximize the total volume of the packed boxes. In the second phase, we apply a genetic algorithm on found solution in previous phase in order to maximize its weight distribution. We use a well-known benchmark test to compare our results with other approaches, considering that our algorithm is not yet completely implemented. This paper also presents a case study of our implementation using some real data in a factory of stoves and refrigerators in Brazil. The obtained results are better than the found results by the factory's system, in reduced time.

Keywords: Container Loading Problem, Weight Distribution, Metaheuristics, Integer Programming, Backtracking, Genetic Algorithms.

1 Introduction

In recent years, many works have approached the Container Loading Problem (CLP), in other words, the task of packing boxes of various sizes within a container optimizing a criterion such as the total loaded volume in a single container or the quantity of used containers. However, relatively few have held the weight distribution constraint.

This work combines heuristics backtracking and Genetic Algorithm (GA) to solve the problem of how to pack the maximum volume of boxes with

Luiz Jonatã Pires de Araújo · Plácido Pinheiro
University of Fortaleza (UNIFOR) - Master Course in Applied Computer Sciences
Av. Washington Soares, 1321 - Bl J Sl 30 - 60.811-905 - Fortaleza - Brazil
e-mail: ljonata@gmail.com, placido@unifor.br

E. Corchado et al. (Eds.): SOCO 2010, AISC 73, pp. 95–102.

dimensions (l_i, w_i, h_i), and quantity b_i, $i = 1,...,m$, into a single container with dimensions (L, W, H), according to space constraints (not overlapping) and weight limit. Therefore, we are concerned with the maximization of the weight distribution, which greatly increases the already high complexity of the problem. An instance of a CLP is ranked as the diversity of box types that can be loaded into the container. According to [9, 19, 20], the cargo can be weakly heterogeneous (many items of relatively few different box types) or strongly heterogeneous (many items of many different box types).

This work is organized as follows. We present in Section 2 some related works. In Section 3, we show our methodology. In Section 4, we list the computational results, and in the end, we make some considerations that include future development.

2 Related Works

The CLP is a kind of the cutting and packing problems, cf. typology introduced on [9]. It is admittedly a NP-hard problem. Theres no known algorithm that resolves it.

However, we can describe a CLP by a integer programming model, such in [7]. This model has a lot of constraints or variables, about $O(n^2)$, being impracticable to apply a solver to directly solve it.On [7] this approach is applied to solve problems with 6 boxes at most. Other work is [15], that presents an exact method, using branch-and-bound, for solving CLP instances with 90 boxes at most.

In the face of exact methods impracticality, many works adopt strategies, heuristics, to avoid direct application. Several heuristics have been proposed in the literature. One of them is the building of box blocks, called layers, vertical or horizontal [5, 13], in which the dimensions of a layer are determined by the first box in the block. Another relevant work is [19] that uses a tree search to decide the size of the layers. Are depicted in [16] guillotine cuts to split the empty space in the container and to fill it.

In addition to the proposed heuristics, there are those which use metaheuristics such as Tabu Search [6], or Genetic Algorithm (GA). For example [10], that built stacks to the top of the container. Then, applying a GA to determine the order of placement of stacks and choose the configuration that results in better use of container.

We increasingly find proposals that seek to combine the best of the exact method with heuristic methods. These proposals are called hybrid methods. A successful work is [17], that generates reduced instances of the problem using genetic algorithms and solve these subproblems by linear programming. This proposal achieved an average allocation between 92 and 95%, but in very large runtime. Other work is [14] that combines parallel tabu with simulated annealing.

There is not a single approach that is the best for all problem types. Each one of the heuristics is better or worse in specific cases or instances of problems.

3 Methodology

Here we present the works that have inspired our methodology. The first is [18], which makes an allocation of layers coordinated through a tree search in order to determine the best depth or width of a layer. The backtracking algorithm navigates through the tree search to determine the best size for each layer. The second work is [8], that creates multiple vertical layers. Layer sets are called segments, which can be rotated or have positions interchanged in order to greedily improve the weight distribution of the container.

The algorithm proposed in this material is divided into two phases. Initially we try to maximize the volume. Then we attempt to maximize the weight distribution of the cargo.

3.1 Phase 1 – Heuristics Backtracking (Maximizing the Volume)

Data Structure. The algorithm builds a tree that consists of nodes, which helps to control the processing flow. The root node receives the input problem of the algorithm, that is to say, a definition of the empty space inside the container and a list of available boxes for packaging. The node tests a strategy to do some packing. If it can not, it changes the strategy. The result of the successful application of a strategy is a list of packed boxes and an output problem, i.e., the definition of a residual space to be filled and a new list of boxes. This output problem is the input problem to a new node in the tree, hierarchically below the node that preceded it. This process continues until we can no longer pack any box, regardless of tested strategy. We show two examples of resolution found to the same problem in Fig. 1.

Another data structure is the box type, set of boxes with the same characteristics. Each box type is associated with a constant, which we call relevance, used to sort

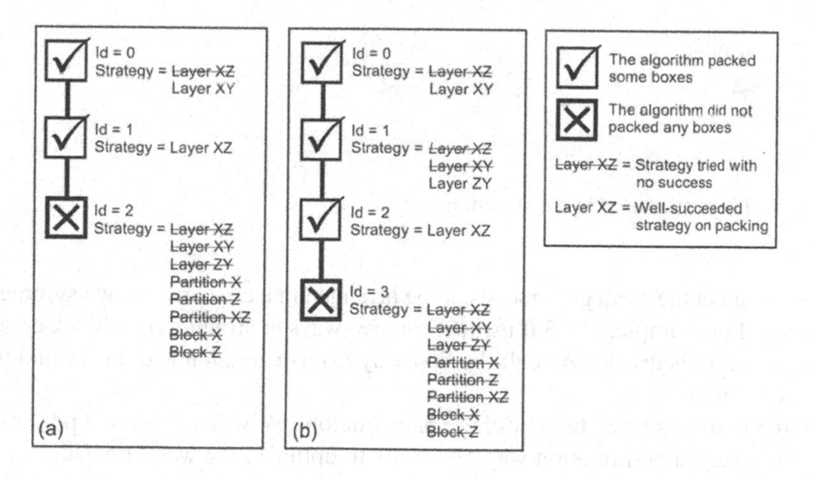

Fig. 1 (a) First found solution. (b) Better solution for the same problem

the box types according to their volume. The higher the volume, the greater the relevance. This constant will allow us to prioritize the large packing boxes, leaving the smaller boxes (easier to be allocated) to the end of the process, when residual space is small.

Heuristics. We saw in Section 2 some popular heuristics (e.g. construction of layers, blocks or stacks) that are better or worse for certain problem types. In our proposal, we formulate a mathematical model in integer linear programming for each heuristic. Each of these models is about $O(n)$ variables and about the same amount of restrictions. Therefore, the application of a heuristic on an instance of the problem shows up extremely fast.

The heuristics used in the algorithm are: Layer XZ (a), Layer XY (b), Layer ZY (c), Partition on X (d), Partition on Z (e), Partition on XZ - Stack (f), Block on X (g) and Block on Z (h). They are all illustrated in Fig. 2.

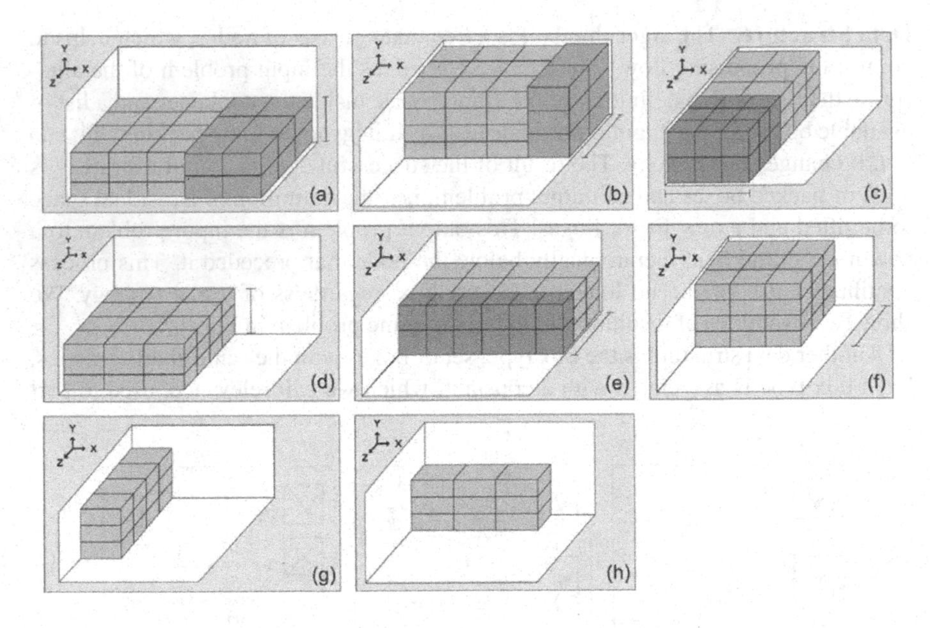

Fig. 2 The heuristics used in the algorithm

Each one of the strategies above can be arranged in a container in two symmetrical ways. For example, Fig. 3 illustrates the two ways to arrange a box block created by 'Layer XZ' heuristic. We call the first way (a) pair rotation and the second way (b) odd rotation.

In the first phase we always use the pair rotation. We will see that the possibility to pack boxes in odd rotation way allows us to optimize the weight balance of the cargo.

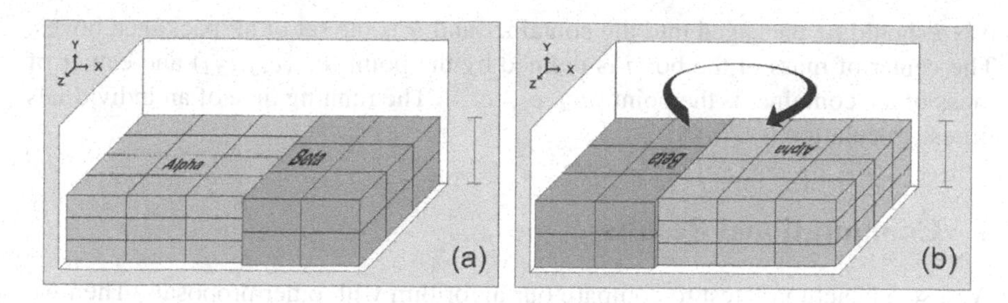

Fig. 3 Two ways of rotation: (a) pair and (b) odd

Backtracking. In Fig. 1-a, we could notice that in the second node the algorithm was successful using the first heuristic. But the heuristic that would lead to a better use of the container at a later node, in that specific case, was the third heuristic, as illustrated in Fig. 1-b. Therefore, a bad solution may be due to a bad choice of a heuristic in the previous node. We use backtracking heuristics to find the best solution though tree states.

3.2 Phase 2 – Genetic Algorithm (Maximizing Weight Distribution)

After the first phase, we obtain a tree with n nodes that corresponds to a packing plan. Given a tree, we can represent it by a binary chromosome with n alleles. An allele i with value equal to 0 indicates that the packed box block in node i, by one of aforementioned heuristics, should be arranged in the way called pair (see Section 3.1). If the value is equal to 1, in the way called odd. Two different chromosomes, obtained from a random solution, are illustrated in Fig. 4. We can notice that in fifth node they differ in rotation. Over the chromosomes, we can apply mutation, crossing over and selection operators.

For each individual we can apply an evaluation function that informs us how well the distribution weight is. We will use a particular formulation: $f = [\sum_{i \in B} p_i(c_{x,i} - c_{x,c})]^2 + [\sum_{i \in B} p_i(c_{z,i} - c_{z,c})]^2$, where p_i is the binary variable that indicates if the

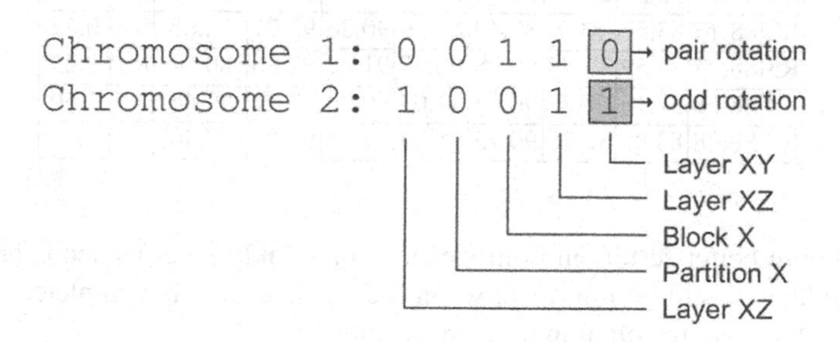

Fig. 4 Different chromosomes obtained from a same solution

box i should be packaged into the container and B is the set of all packaged boxes. The center of mass of the box i is defined by the point $(c_{x,i}, c_{y,i}, c_{z,i})$ and center of mass of the container is the point $(c_{x,c}, c_{y,c}, c_{z,c})$. The running time of an individuals fitness calculation is $O(n)$.

4 Computational Results

We use a benchmark test to compare our algorithm with other proposals. Then we present a case study in which we compare our results with those of the company and another resolution.

The results were obtained using an Intel Core 2 Duo 2.1 GHz with 3 GB of RAM. The operating system is Windows Vista Home Edition. The development platform is Java 6.0 and Eclipse 3.1.1 tool. The used solver was CPLEX 9.0.

It is important to consider that we have not implemented the heuristics backtracking yet. The algorithm just returns the first solution. We hope to significantly increase percentage of use of the container. We havent implemented the genetic algorithm for maximizing the balance of the weight of the load too. So we dont show results about this.

Benchmarking problems. We have used the collection of test data sets for a variety of Operational Research problems, originally described in [1]. These data sets contain several types of problems, from homogeneous to strongly heterogeneous. The library is divided into files named BR1 to BR5 and we compare the average of use of the container space of our algorithm (represented by HBGA) with: a genetic algorithm proposed in [10] (GA), the constructive algorithm on [4](CA), a constructive algorithm by Bischoff [3](HBal), the hybrid genetic algorithm described in [5](HGA), the parallel genetic algorithm in [11](PGA), a proposed heuristic in [2](PH) and Tabu Search proposed in [6](TS). We also present our standard deviation σ and the average execution time in seconds.

Table 1 Comparing some proposals

File	GA	CA	HBal	HGA	PGA	PH	TS	HBGA	σ	Time
BR1	86,77	83,37	81,76	87,81	88,10	89,39	93,23	**84,46**	6,2	1,35
BR2	88,12	83,57	81,7	89,40	89,56	90,26	93,27	82,66	8,4	0,72
BR3	88,87	83,59	82,98	90,48	90,77	91,08	92,86	80,54	8,1	1,25
BR4	88,68	84,16	82,6	90,63	91,03	90,90	92,40	79,46	7,5	2,50
BR5	88,78	83,89	82,76	90,73	91,23	91,05	91,61	77,40	7,4	1,71

We observe better results in homogeneous or weakly heterogeneous problems. These results will still be improved when the backtracking is completed. We emphasize the low average running time, in seconds.

Case study. ESMALTEC is a highly regarded Brazilian company, founded on 1963, that produces stoves, refrigerators, freezers and water coolers. It is one of the most modern factories in Latin America, and sells its products in over 50 countries.

Containers are used to transport the many type of products. Real data about the most common cases were used by [17] to compare the companys results with results of the Hybrid Algorithm (HA). We add a work [8] that lists some results of problems with a single type of box. The work cites [12] (represented by GMM in Table 2), the layering approach of Bischff [4](LA) and the column building procudure of Bischoff and Ratcliff [4](CB). The values are approximate (indicated in Table 2 by *) because they are presented in graphic in [8]. Our proposal is represented by HBGA.

Table 2 Real data in the factory

Problem	ESMALTEC	AH	GMM*	CB*	LA*	HBGA	Time (sec)
1	80,8	93,2	79,8	82,8	85,9	**84,81**	0,450
2	76,5	76,5	79,8	82,8	85,9	67,89	0,062
3	93,7	96,6	79,8	82,8	85,9	**94,80**	0,357
4	95,6	95,6	79,8	82,8	85,9	**87,06**	0,186
5	91,2	95,4	79,8	82,8	85,9	**95,37**	0,218
6	84,1	87,8	79,8	82,8	85,9	72,88	0,108

We can notice that our implementation, even incomplete, had a better percentage than the majority of the proposals on problems 1, 3, 4 and 5. In most cases, the results are better than the results of the company. Again, we emphasize the low average running time, about 0.23 seconds.

5 Final Remarks

The current algorithm version finds only the first solution, which hardly ever is the best solution that maximizes the loaded volume. Even so, we have found results, in some cases, better than the others approaches. Mainly whe the cargo is homogeneous. We intend to implement the heuristics backtracking which allow us to improve the use of the container.

We also intend to implement the second phase of the proposal, the GA for the weight distribution, and to show the results.

Finally, we expect to use the implemented algorithm and its results in our case study, suggesting better results to ESMALTEC company.

Acknowledgements. The author Placido Rogerio Pinheiro is thankful to the National Counsel of Technological and Scientific Development (CNPq) for the support received on this project.

References

1. Beasley, J.E.: OR-Library: distributing test problems by electronic mail. Journal of the Operational Research Society 41(11), 1069–1072 (1990)
2. Bischoff, E.E.: Three dimensional packing of items with limited load bearing strength. European Journal of Operational Research 168, 952–966 (2006)
3. Bischoff, E.E., Janetz, F., Ratcliff, M.S.W.: Loading Pallets with Nonidentical Items. European Journal of Operational Research 84, 681–692 (1995)
4. Bischoff, E.E., Ratcliff, M.S.W.: Issues in the Development of Approaches to Container Loading. Omega 23, 377–390 (1995)
5. Bortfeldt, A., Gehring, H.: A Hybrid Genetic Algorithm for the Container Loading Problem. European Journal of Operational Research 131, 143–161 (2001)
6. Bortfeldt, A., Gehring, H., Mack, D.: A Parallel Tabu Search Algorithm for Solving the Container Loading Problem. Parallel Computing 29, 641–662 (2002)
7. Chen, C.S., Lee, S.M., Shen, Q.S.: An analytical model for the container loading problem. European Journal of Operations Research 80, 68–76 (1993)
8. Davies, A.P., Bischoff, E.E.: Weight distribution considerations in container loading. European Journal of Operations Research 114, 509–527 (1999)
9. Dyckhoff, H.: A typology of cutting and packing problems. European Journal of Operational Research 44, 145–159 (1990)
10. Gehring, H., Bortfeldt, A.: A Genetic Algorithm for Solving the Container Loading Problem. Internat. Trans. Internat. Trans. in Operational Research 4, 401–418 (1997)
11. Gehring, H., Bortfeldt, A.: A Parallel Genetic Algorithm for Solving the Container Loading Problem. International Transactions in Operational Research 9, 497–511 (2002)
12. Gehring, H., Menschner, K., Meyer, M.: A computer-based heuristic for packing pooled shipment containers. European Journal of Operational Research 44, 277–288 (1990)
13. George, J.A., Robinson, D.F.: A heuristic for packing boxes into a container. Computers and Operations Research 7, 147–156 (1980)
14. Mack, D., Bortfeldt, A., Gehring, H.: A parallel hybrid local search algorihtm for the container loading problem. International Transactions in Operations Research 11, 511–533 (2004)
15. Martello, S., Pisinger, D., Vigo, D.: The three-dimensional bin packing problem. Operational Research 48, 256–267 (2000)
16. Morabito, R., Arenales, M.: An and/or-graph approach to the container loading problem. International Transactions in Operational Research 1(1), 59–73 (1994)
17. Nepomuceno, N., Pinheiro, P.R., Coelho, A.L.V.: Tackling the Container Loading Problem: A Hybrid Approach Based on Integer Linear Programming and Genetic Algorithms. In: Cotta, C., van Hemert, J. (eds.) EvoCOP 2007. LNCS, vol. 4446, pp. 154–165. Springer, Heidelberg (2007)
18. Pisinger, D.: Heuristc for the Container Loading Problem. European Journal of Operational Research 141, 382–392 (2000)
19. Pisinger, D.: Heuristics for the Container Loading Problem. European Journal of Operational Research 141, 143–153 (2002)
20. Wascher, G., Hausner, H., Schumann, H.: An improved typology of cutting and packing problems. European Journal of Operational Research 183(3), 1109–1130 (2007)

A Decision Support System for Logistics Operations

María D. R-Moreno, David Camacho, David F. Barrero, and Miguel Gutierrez

Abstract. This paper describes an Artificial Intelligence based application for a logistic company that solves the problem of grouping by zones the packages that have to be delivered and propose the routes that the drivers should follow. The tool combines from the one hand, Case-Based Reasoning techniques to separate and learn the most frequent areas or zones that the experienced logistic operators do. These techniques allow the company to separate the daily incidents that generate noise in the routes, from the decision made based on the knowledge of the route. From the other hand, we have used Evolutionary Computation to plan optimal routes from the learning areas and evaluate those routes. The application allows the users to decide under what parameters (i.e. distance, time, etc) the route should be optimized.

1 Introduction

The actual demand of precise and trustable information in the logistic sector has driven the development of complex Geographic Information Systems (GIS) very useful for planning their routes. Those systems are combined with Global Position Systems (GPS) for positioning the different elements involved in the shipping. However, those systems although very useful, cannot take decisions based on, for example shortcuts in the route that human operators learn from the experience.

David F. Barrero · María D. R-Moreno
Departmento de Automática. Universidad de Alcalá, Madrid, Spain
e-mail: (david,mdolores)@aut.uah.es

David Camacho
Departamento de Informática. Universidad Autónoma de Madrid, Madrid, Spain
e-mail: david.camacho@uam.es

Miguel Gutierrez
Espi & Le Barbier
e-mail: miguel.gutierrez@espilebarbier.com

E. Corchado et al. (Eds.): SOCO 2010, AISC 73, pp. 103–110.

In our particular problem, the logistic company has a set of logistic distributors with General Packet Radio Service (GPRS) connexions and Personal Digital Assistants (PDAs) (around 1200 in the whole Spanish geography) where the list of shippings is stored for each day. The order in which they should be carried out is decided by the distributor depending on the situation: density of the traffic, state of the highway, hour of the day and place of the shipping. For example, it is better to ship in industrial areas early in the morning if the traffic is less dense. The list of tasks for each distributor is supplied each day from a central server.

Within this context, the company needs a decision support tool that, from the one hand allows them to decide the best drivers behaviors and learn from them. And from the other hand, to plan the routes and evaluate and compared them under different parameters.

There are several available tools that address (partially) the described problem. For example, the ILOG SOLVER provides scheduling solutions in some domains such as manufacturing or transportation and the ILOG DISPATCHER provides extensions for vehicle routing. STRATOVISION is a Decision Support and Modeling System for Logistics Strategy Planning that allows the user to represent a scenario and interact with it in an easy way. AIMSUN allows evaluating different transportation solutions. But, none of these tools combine what the experience operators do (learning from their decisions), and plan optimal routes from the learnt knowledge.

This paper presents a tool that combines Case-Based Reasoning (CBR) to learn humans decisions and Evolutionary Computation (EC) to plan for optimal routes. The structure of the paper is as follows. Section 2 presents the different subsystems that the application is subdivided into. Next, section 3 describes in detail the CBR architecture and the algorithms used. After, the GA and the parameters used for the route optimization are introduced in section 4. Then, section 5 shows an experimental evaluation of the application with the real data provided by the logistic company. Finally conclusions and future work are outlined.

2 Application Architecture

The application is subdivided in four subsystems:

- The Data Processing Subsystem: takes the data from the files provided by the logistic company in CSV format (Comma Separated Values). It analyses that the files are valid, loads the information in data structures and performs a first statistic analysis of each of the drivers contained in the file such as the number of working days, number of physical acts, average of physical acts performed, or average of the type of confirmation received by the client.
- The Loading Data Subsystem: is in charge of obtaining and loading the geographic information (latitude and longitude) of each address and the distances among addresses. This information is provided by mean of the Google Maps API. During the process of calculating the coordinates, we group the same addresses (a driver can ship several packages in the same address) for a given date and a driver into one. We name that *act*. So in the database, the addresses will

be loaded as *acts* and an extra field will be added to represent the number of deliveries and pick ups for that act.

- The Learning Working Areas (LWA) Subsystem: creates zones or working areas for each driver using the data loaded and processed in the previous subsystems. To generate that subdivision we have used the zip code as the baseline. So, a zone or a working area can be represented by a zip code or more than one. It also allows us to visualize (using the Google Maps API) the different working areas.
- The Learning Manager Task (LMT) Subsystem: plans for routes in a date selected by the user. We can specify the number of different plans we want as output and compare them with the original plan performed by the driver. In the comparison and the generation of plans, we can use different parameters to measure the quality of the plans such as the total distance, positive rate of LWA, time, etc. The planning algorithm can adapt its answer to the driver behavior and generate plans according to it, if that is what we want.

3 Case Based Reasoning Architecture

Cased Based Reasoning (CBR) [1] is an Artificial Intelligence (AI) technique that solves new problems based on the acquired knowledge about how similar problems were solved in the past. Any CBR system can be defined using two main elements: the *cases* or concepts which are used to represent the problem, and the knowledge base that is used to store and retrieve the cases. To handle this knowledge the CBR systems needs to define several processes to *Retrieve* the most similar case(s) in the knowledge base, *Reuse* the selected case using its solution adapted to the current problem, *Revise* the adapted solution to verify if it can solve the actual case (if not the process starts again), and finally, *Retain* the solution if it is satisfactory.

In our case, the knowledge base is extracted from the drivers experience and later it is reused for optimization and logistic issues. This knowledge is given by the files provided by the company (see the Data Processing Subsystem section).

The information loaded in the database can be used it straightforward (i.e. address and time delivery can be used to estimate how much time is necessary to complete a particular ship), or it can be used to indirectly learn from the experience drivers. The drivers know which areas have thick traffic so they can take alternative paths to reduce time and gas consumption. During the working day, the driver may stop in places not related to the shipping to rest or eat. So this information has also some disadvantages, mainly caused by the deficiency and irregularity of the data given by the company. The information has a lot of noise: there are many mistakes in the zip codes and the street names that are hard to correct. The shipping time that could be used to calculate the time between some points in the path, it is not reliable since sometimes the drivers annotate the hour after they have done some deliveries.

So, our CBR algorithm uses this information in a simplified way to prevent that the noisy information could affect to the optimization process. The generation of the final CBR information, such as the working areas of the drivers, are carried out using statistical considerations (average of behaviours followed by all the drivers analysed) to minimize the noise.

Then, the following step in our CBR algorithm is to group the shippings. We can define the concept of the *working zone* or *working area* (WA). It represents a particular zone in a city, or in a country, where one or several drivers will work shipping objects (as mentioned before, we have called them **acts**). Usually the logistic company define (using a human expert) these WA and assign them to a set of drivers. Minimizing the overlapping of these areas among different drivers is essential to minimize the number of drivers and the deliver time for a set of acts. The automatic generation of these working areas is the target of our CBR algorithm, these WA will be used later in the optimization process.

The algorithm starts defining a grid that represents all the available postal codes ($Z_i, i \in [1..n]$) for a particular city (or county). These postal codes are used to fix a geographical centroid so we can calculate (using any standard algorithm, i.e. based on GPS values) the distance between two postal codes Z_i and Z_j ($distZ_i, Z_j$). On the other hand, the information from driver's log is used to calculate how many acts belongs to the same postal code, and how many acts occurs between adjacents postal codes (a driver could work in a particular postal code, or could work in several postal codes). This information is represented using a parameter s_i which is calculated as $s_i = \sum(acts(Z_i \rightarrow Z_j) + acts(Z_j \rightarrow Z_i))$, this parameter represents the jumps between two postal codes (we can expect that if several acts are interleaved, and they are placed in different postal codes, these should be enough closer to maintain the shipping costs low).

Let us now consider a particular driver's log as a list of m ordered acts ($act_k, k \in [1..m]$). A new value $\theta = \sum(d_k/s_k)$, is calculated for each log, where
$d_k = dist(act_k(Z_k), act_{k+1}(Z_{k+1}))$.

Using the postal codes stored in each log we generate a set of LWA using a clustering algorithm based on the value of θ and δ [1]. The algorithm works as follows, using the drivers available (and the predefined value of δ) all the acts are grouped into a set of clusters, then these clusters are compared among drivers: If a particular cluster is detected in different drivers, it is given to the system as a new learned working area. The quality of the learned clusters will depend on the number of drivers that has this cluster, so the system can take different actions in the planning process taking into account how good this cluster is. Currently, we consider that any learned cluster that belongs at least to three different drivers has enough quality to be used directly by the system. This could be modified in the near future and allow the user to modify the planning process using a reliability factor of these learned clusters.

Finally, each LWA learned is stored as a new case in the data base, for each driver the algorithm is applied and the LWA zones are generated.

[1] The value of δ was obtained from an empirical evaluation of several drivers and working days randomly selected. This value was selected analysing the LWA_1 generated in the first execution of the algorithm.

4 Route Optimization

One of the main features of the proposed system is its capacity to suggest efficient routes. The term eficient in this context should be interpreted as relative to a set of routes designed by a human expert and provided to the system as input. So, the goal of the described system is to improve a given set of routes rather than generate complete new routes. This characteristic is used by the optimization engine to guide its search. The definition of the zones is provided by the CBR described in the previous section so it doesn't have to handle this task and thus, the route optimization can be considered as a variation of the Travel Salesman Problem (TSP). The selection of the optimization algorithm is critical to the success of the system. Evolutionary Computation (EC) [3] provides a set of tools that can be used within this scenario.

EC is a collection of algorithms inspired in the biological evolution. Regardless of the flavour of the specific EC technique, they share three characteristics, (1) they use a population of individuals that represent each one a solution in the search space, (2) individuals are modified using a genetic operator, and (3), individuals are under a selective pressure. The result is a evolution of the population that would eventually converge to a global solution. From an AI point of view, EC is a set of stochastic search algorithms. Depending on how individuals are represented and, how they are modified, we can find several algorithms, one of the most sucessful ones are Genetic Algorithms (GA) [4].

The GA implemented is based on the work of Sengoku described in [5]. This GA encodes the solution using a classical permutation codification [2], where the acts are coded as integer numbers between one and the number of acts in a fixed length chromosome. Since no act can be visited twice, no integer in the chromosome is allowed to be repeated. Indeed the valuable information is not the presence of the integer but rather information is kept in the adjacency. In this way, the chromosome $A_1 = \{4,3,2,5,6,1\}$ represents the path $4 \rightarrow 3 \rightarrow 2 \rightarrow 5 \rightarrow 6 \rightarrow 1$, and it is equivalent to another chromosome $A_2 = \{5,6,1,4,3,2\}$ representing $5 \rightarrow 6 \rightarrow 1 \rightarrow 4 \rightarrow 3 \rightarrow 2$. The route that they code is the same because the genes have the same adjacency.

GAs require a mechanism to evaluate the quality of individuals, or fitness. This is a key subject in any GA design that may determine its sucess or failure. The fitness function has a close relationship with the value that the solution provides to the user. When optimizing routes, the fitness that an individual scores provide a refference of time or distance savings. In our case, we have used an aggregate scalar fitness function which evaluates the individual based on different characteristics.

A human made route is given to the GA as reference as well as a classification of act in zones. The fitness evaluation is done comparing the solution to this reference route. Of course, a route can outperform or not another route depending on the criteria that is applied. Our system uses four criteria or qualifications: Time, Distance, Zone and Act.

Using these qualifications we can evaluate different aspects related to the route, such as time or distance. In order to provide a syntetic estimation of the quality of the individual, these qualifications are aggregated in a linear combination that conforms the fitness function, as is expressed in equation 1.

$$f(A) = 0.4\omega_d\Delta_d(A) + 0.4\omega_t\Delta_t(A) + \omega_z\Delta_z(A) + \omega_a\Delta_a(A) \quad (1)$$

where ω_i are coeffients associated to distance (ω_d), time (ω_t), zones (ω_z) and acts (ω_a). These coefficients are used to weight the contribution of each category to the fitness. It is possible to change the priority of qualifications that the user prefer to optimize just modifing the coefficients ω_i. The function $\Delta_i(A)$ returns the relative difference between the route codified in the chromosome A and the reference route for each one of the described categories. By default, ω_d and ω_t are set to 0.35 while ω_z and ω_a are both set to 0.15.

The TSP is a problem where it is not possible to know when a global maximum is achieved, so a convergence criteria is required to stop the execution of the GA. In this case, the GA is supposed to have converged if the average fitness in generation i is less than 1% better than the fitness in generation $i-1$. The initial population is generated randomly avoiding repeated individuals.

Our GA uses the same evolution strategy than [5]. Given a population M of routes in generation i, the N best routes are cloned. In an attempt to avoid a premature convergence due to the loose of genetic diversity, the routes are sorted by their fitness value and the adjacent one are compared. Those routes whose fitness have a difference less than ε are removed. To maintain constant the number of individuals in each generation, $M-N$ individuals are generated by means of a Greedy Subtour Crossover [5] where the parents are radomly selected. Then a mutation operator called 2opt is applied with a probability p_c. This operator swaps two random points in the route if and only if the new individual is fitter than the old one. In case that 2opt did not generate a fitter individual it is not modified.

5 Experimental Results

One of the main problems we encountered in testing was the high percentage of mistakes in the addresses in the input files. Without any pre-processing the number of errors in the streets or zip codes is around 35%. After some preprocessing (such as using the address to update incorrect zip codes, or eliminating or adding some characters and searching again) the percentage drops to 20%. This is still very high when attempting to perform an automatic comparison of the routes followed by the drivers and the ones generated by our tool.

In our first attempt to compare the results, we manually cleaned the input files of 10 drivers during one month. The drivers choosen average 25 to 40 acts each day. The improvement obtained on average by our tool is 26% if we use the distance as the comparison parameter and 20% if we use the time.

But these results require of some explanations using a *typical* driver of the set analised previously. By typical we mean that his behaviour is normal and the route generated does not contain too much noise (i.e. a noisy route would perform in 30 mn half of the shippings).

The comparison of the results is carried out on the fitness of the best individual obtained after the execution of the algorithm compared to the fitness of the original itinerary on the acts of one day. Table 1 shows the values of the fitness in both itenaries and the values of the different parameters that are in the fitness.

Table 1 Values of the fitness in the original and planned routes

Route	Fitness value	Distance (kms)	Distance improv. (%)	Time (minutes)	Time improv. (%)	Zone calification	Act calification
Original	2.839	140.7	0.0	209.8	0.0	0.892	1.0
Planned	10.75	97.0	31.04	149.5	28.75	1.0	0.585

Comparing the original fitness (2.839) with the planned one (10.75) does not offer much information until we analyse the contributions of the different parameters. The number of kms varies from 140.7 Km in the original itinerary to 97 Km in the planned one. This provides a distance improvement of the 31.04%.

The information related to the time is extracted from the columns time (min) and time improvement in (%). The driver took 209.8 minutes to perform all the shippings while our algorithm took 149.5 minutes. This means a reduction of 28.75% of the employed time.

Analazing graphically the results, a similar behavior is observed with the time and the distance in both routes. At the beginning, the shippings performed by the driver takes a big advantage over the planned ones (over 30 kms or 60 minutes). But in the last part of the shippings the planned route beats the original plan. This is a common behaviour in all the drivers analysed: at the beginning the humans try to do the shippings closer to the starting point. Instead, the EC algorithm looks for a solution that does not minimize the initial part of the route but all of it. Figure 1 shows the original and planned routes using the Google Maps API. Another common mistake that we have detected is that the drivers try to do the shippings that

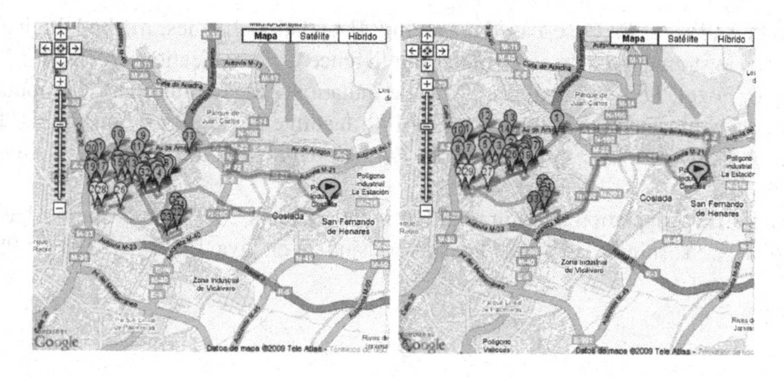

Fig. 1 Plans comparison using Google Maps

are in the same street. Although it seems a logical reasoning, some streets cross the city from north to south or east to west. Following that criteria can higly increase the number of kms since the driver has to drive back to a point close to the previous shipping.

The time for generating the results by our EC algorithm is not crucial since planning routes can be generated off-line by our tool and given to the driver before he starts his working day. The logistic company distributes the packages of each driver in advance and just a small percentage of new unknown pick ups occur when the driver has already started the shipping.

6 Conclusions and Future Work

In this paper we have described the AI-based application that we have developed for a logistic operator company that combines Case-Based Reasoning (CBR) and Evolutionary Computation (EC) techniques. CBR is used to separate and learn the most frequent areas that the experienced drivers follow. These techniques allow one to separate the daily incidents that generate noise in the routes, from the decision made based on the knowledge of the route. The EC techniques plan optimal routes from the learning areas and evaluate those routes.

Our next step will be to include a new module in the application that pre-process automatically or in a mixed-initiative way the addresses bad introduced by the drivers. Due that the 20% of the streets and zip code cannot be automatically corrected, this represent a bottleneck to efficiently show the results of the tool.

Acknowledgements. We want to thank Antonio Montoya for his contribution in the tool developed. This work has been supported by the Espi & Le Barbier company and the public projects funded by the Spanish Ministry of Science and Innovation under the projects COMPUBIODIVE (TIN2007-65989), V-LeaF (TIN2008-02729-E/TIN) and by Castilla-La Mancha project PEII09-0266-6640.

References

1. Aamodt, A., Plaza, E.: Case-based reasoning: Foundational issues, methodological variations, and system approaches. Artificial Intelligence Communications 7(1), 39–52 (1994)
2. Bäck, T., Fogel, D.B., Michalewicz, Z.: Permutations. In: Evolutionary Computation 1. Basic Algorithms and Operators, pp. 139–149. Institute of Physics Publishing (1984)
3. Eiben, A.E., Smith, J.E.: Introduction to Evolutionary Computing. Springer, Heidelberg (2009)
4. Holland, J.H.: Adaptation in natural and artificial systems. MIT Press, Cambridge (1992)
5. Sengoku, H., Yoshihara, I.: A fast tsp solver using ga on java. In: 3rd AROB (1998)

Greenhouse Heat Load Prediction Using a Support Vector Regression Model

João Paulo Coelho, José Boaventura Cunha,
Paulo de Moura Oliveira, and Eduardo Solteiro Pires

Abstract. Modern greenhouse climate controllers are based on models in order to simulate and predict the greenhouse environment behaviour. These models must be able to describe indoor climate process dynamics, which are a function of both the control actions taken and the outside climate. Moreover, if predictive or feedforward control techniques are to be applied, it is necessary to employ models to describe and predict the weather. From all the climate variables, solar radiation is the one with greater impact in the greenhouse heat load. Hence, making good predictions of this physical quantity is of extreme importance. In this paper, the solar radiation is represented as a time-series and a support vector regression model is used to make long term predictions. Results are compared with the ones achieved by using other type of models, both linear and non-linear.

João Paulo Coelho
Instituto Politécnico de Bragança, Campus de Santa Apolónia - 5301-857 Bragança, Portugal,
CITAB - Centro de Investigação e de Tecnologias Agro-Ambientais e Biológicas
e-mail: jpcoelho@ipb.pt

José Boaventura Cunha
Universidade de Trás-os-Montes e Alto Douro, Dep. Engenharias, 5001-801 Vila Real, Portugal
e-mail: jboavent@utad.pt

Paulo de Moura Oliveira
Universidade de Trás-os-Montes e Alto Douro, Dep. Engenharias, 5001-801 Vila Real, Portugal
e-mail: oliveira@utad.pt

Eduardo Solteiro Pires
Universidade de Trás-os-Montes e Alto Douro, Dep. Engenharias, 5001-801 Vila Real, Portugal,
CITAB - Centro de Investigação e de Tecnologias Agro-Ambientais e Biológicas
e-mail: epires@utad.pt

E. Corchado et al. (Eds.): SOCO 2010, AISC 73, pp. 111–117.
springerlink.com

1 Problem Statement

Nowadays, a substantial part of agricultural production takes place in greenhouses, which enables to tune the crop growing by modifying, artificially, the environmental conditions and the plant's nutrition. The main goal is to optimize the balance between the production economic return and the operation costs of the climate actuators. Severe environment and market restrictions, jointly with an increasing tendency of the fuel price, motivate the development of more "intelligent" management and control strategies.

State-of-the-art greenhouse climate controllers are based on models to simulate and predict greenhouse environment behaviour [2] [7]. These models must be able to describe indoor climate process dynamics, which are functions of both control actions taken and outside climate. Moreover, if predictive or feedforward control techniques are to be applied, it is necessary to employ models to describe and predict the outside climate, being the most relevant the air temperature and solar radiation. The latest, it's the exogenous variable which most influences the thermal load during the day, hence the importance of making good forecasts.

The overall objective of this study is to predict the solar radiation future trend by taking into consideration only the time series past observations. In this context, several models were tested with a special focus given to the one obtained by a support vector regression strategy.

2 The Support Vector Regression Model

The Support Vector Machine (SVM) concept, introduced by Vapnik [8] in the first half of the nineties, with proper modification can be used in regression problems. This paradigm, denoted by Support Vector Regression (SVR), has substancial differences regarding the classification oriented approach. Even so, SVM and SVR share the common fact that both require solving a quadratic constrained optimization problem.

In the SVR, the goal is to make a mapping from a d-dimensional input vector $\mathbf{x}$ into a scalar y using an hyperplane. It is important to note that this approximation in not done in the input space but in an alternative space, of higher dimensionality, denoted by feature space. Thus the function approximation problem, in the SVR universe, resumes to find a vector $\mathbf{w}$ and a scalar b in order to meet the following equality:

$$\hat{y} = \mathbf{w}^T \phi(\mathbf{x}) + b \tag{1}$$

where $\phi(\cdot)$ is a characteristic function which expands the input vector $\mathbf{x}$ into a higher dimensionality space.

The problem definition resembles standard least squares. However, the SVR parameters are not estimated by minimizing a quadratic-in-the-error cost function but by the following alternative cost function [8]:

$$J = \max\{0, |y - \hat{y}| - \tau\}, \tau \geq 0 \tag{2}$$

In the above expression, τ is a parameter that defines the width of a dead-band in J. By using function (2) it is possible to have several models with the same cost value. Among this infinite universe of possible models a plausible choice is to select the one which gives the best generalization ability [4] [8]. Within this framework, the model parameters are computed by solving the following (constrained) quadratic optimization problem.

$$\begin{array}{l} \min\limits_{\|\mathbf{w}\|} \frac{1}{2}\mathbf{w}^T\mathbf{w} + \mathbf{C}^T(\xi_1 + \xi_2) \\ s.t.\ \mathbf{y} - (\mathbf{X}^T\mathbf{w} + \mathbf{b}) \leq \tau + \xi_1 \\ \quad (\mathbf{X}^T\mathbf{w} + \mathbf{b}) - \mathbf{y} \leq \tau + \xi_2 \\ \quad \xi_1 \geq \mathbf{0}, \xi_2 \geq \mathbf{0} \end{array} \tag{3}$$

where: $\mathbf{y} = \begin{bmatrix} y_1\ y_2\ \cdots\ y_N \end{bmatrix}^T$, $\mathbf{X} = \begin{bmatrix} \mathbf{x}_1\ \mathbf{x}_2\ \cdots\ \mathbf{x}_n \end{bmatrix}$, $\mathbf{b} = b \cdot \mathbf{1}$, $\mathbf{C} = C \cdot \mathbf{1}^1$ and ξ_1 and ξ_2 are slack vectors, with the same dimension as $\mathbf{y}$, introduced to let some input vectores to stand out of the $[y - \tau, y + \tau]$ interval. The coefficient C defines a commitment between estimation error and maximum margin. In [1] some heuristics to tune the C and τ parameters are presented.

Usually the optimization problem is not solved in the primal space but in the dual one. By building the Lagrangian, and taking into consideration the Karush-Kuhn-Tucker conditions [5], the optimization problem in the dual form can be represented by:

$$\begin{array}{l} \max\limits_{\alpha_1,\alpha_2} -\frac{1}{2}(\alpha_1 - \alpha_2)^T \mathbf{X}^T\mathbf{X}(\alpha_1 - \alpha_2) + (\alpha_1^T - \alpha_2^T)\mathbf{y} - (\alpha_1^T + \alpha_2^T)\tau \\ s.t.\ (\alpha_1 - \alpha_2)^T \mathbf{1} = 0 \\ \quad \mathbf{0} \leq \alpha_1 \leq \mathbf{C} \\ \quad \mathbf{0} \leq \alpha_2 \leq \mathbf{C} \end{array} \tag{4}$$

where α_1 and α_2 are the Lagrange multipliers.

From the solution of this problem it's possible to state the hyperplane equation by using the following formulation:

$$\hat{y} = (\alpha_1 - \alpha_2)^T \mathbf{X}^T\mathbf{X} + b \tag{5}$$

Note that the bias term b can be computed in several ways [4] such as by using the $(y, \mathbf{x})$ pairs for which the respective Lagrange multiplier is positive.

Usually a simple linear model is unable to appropriately adjust the training data. However, it's possible to extend the previous problem to cope with situations in which the dependencies are non-linear. This can be accomplished by introducing the concept of kernel function [8]. The kernel is a function that represents the dot product of the input space vectors into some other characteristic space, i.e. $\phi(\mathbf{x}_i)^T \phi(\mathbf{x}_j) = K(\mathbf{x}_i, \mathbf{x}_j)$. The choice of this function is tricky. The use of Gaussian or polynomial kernels seems to be a reasonable choice for the majority of the

[1] $\mathbf{1}$ refers to a column vector of ones with dimension N in which the later stands for the number of training examples.

problems [4]. By appling the kernel concept to the problem defined by (4) it is possible to obtain (6).

$$
\begin{aligned}
&\max_{\alpha_1,\alpha_2} -\tfrac{1}{2}(\alpha_1-\alpha_2)^T K(\mathbf{X},\mathbf{X})(\alpha_1-\alpha_2)+(\alpha_1^T-\alpha_2^T)\,\mathbf{y}-(\alpha_1^T+\alpha_2^T)\,\tau \\
&s.t.\ (\alpha_1-\alpha_2)^T\mathbf{1}=0 \\
&\qquad \mathbf{0}\le\alpha_1\le\mathbf{C} \\
&\qquad \mathbf{0}\le\alpha_2\le\mathbf{C}
\end{aligned}
\tag{6}
$$

3 Solar Radiation Prediction Results

Solar radiation, represented as a time-series, is a very complex prediction problem due to the extreme uncorrelated high frequency components. Although the low-frequency profile could be obtained from a deterministic radiation model, the high-frequency oscillation due to disturbances, such as clouds and atmosphere attenuation, is extremely difficult to predict by using only past information.

In this section, a comparative study of several types of models will be investigated concerning their ability to predict the solar radiation in a sixty steps ahead horizon. The data used was acquired in two consecutive days using a one minute sampling period. It is assumed that the dynamic system state space is not directly observable. Hence one will try to capture it's behaviour by fitting it in a time-series structure. The techniques employed for data modeling were:

1. A ten pole filter with coefficients computed, using the first day data, by means of a least squares algorithm.
2. A 10th order AR model with coefficients computed iteratively using the recursive least squares.
3. A feedforward time-delayed Neural Network (ANN), with one hidden layer and dimension ten. The network structure involves one layer with five hyperbolic tangent neurons. The training phase was accomplished by minimizing the estimation error using the *Levenberg-Marquardt* optimization algorithm.
4. A Neuro-Wavelet structure as discussed in [3]. This architecture employs five ANN each one predicting a filtered version of the original time-series.
5. Two support vector regression models: one using a linear kernel and the other using a *Gaussian* kernel.

Table 1 and figure 1 presents the results obtained by applying each of the above enumerated strategies using the second day as validation data. The figures-of-merit used were: the prediction error trajectory (PI_1) and the percentage of change in direction (PI_2). The former is computed using equation (7) and the latest by using the expression (8).

$$
PI_1(N,h) = \frac{1}{N}\sum_{n=0}^{N-h-1}\left\{\frac{1}{h}\sqrt{\sum_{k=0}^{h-1}\left(y[k+n]-y_p[k+n|n]\right)}\right\}
\tag{7}
$$

$$PI_2(N,h) = \frac{1}{N} \sum_{n=1}^{N-h-1} \eta(n) \tag{8}$$

In expression (8) N refers to the number of observations, h to the prediction horizon and $y_p[k+n|n]$ is the k-step ahead predicted output starting from $k=n$. In the performance index PI_2, the function $\eta(n)$ is computed using,

$$\eta(n) = \sum_{k=n+1}^{n+h} u(\Delta y[k] \cdot \Delta y_p[k|n]) \tag{9}$$

where: $\Delta y[k] = y[k] - y[k-1]$, $\Delta y_p[k|n] = y_p[k|n] - y_p[k-1|n]$ and $u(\cdot)$ refers to the *Heaviside* (or discrete step) function.

The results reveal better performance for non-linear models when compared to linear ones. But it's complexity is, by far, larger and more sensitive to several tuning parameters. We remark that the performance of the SVR, in this particular problem, was far from expected, especially when compared to the results obtained by [6] in artificial time-series.

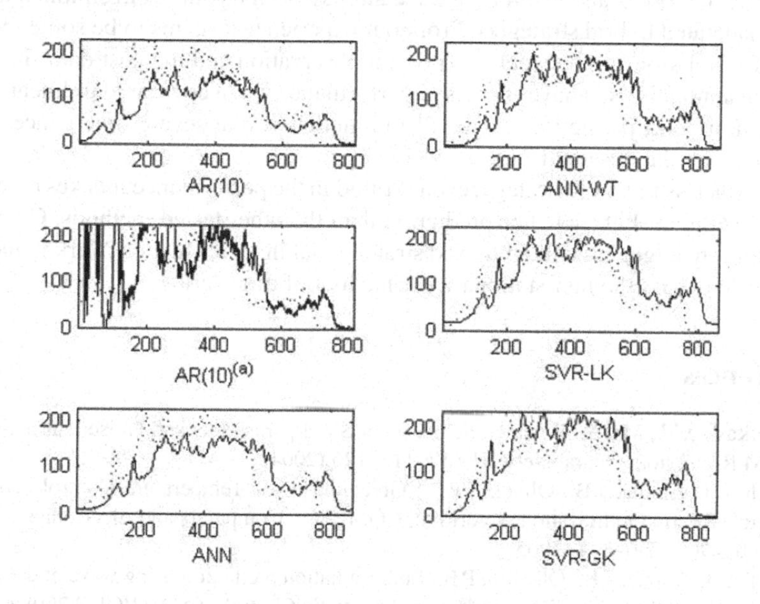

Fig. 1 Measured versus predicted results. In the figure the measured value is represented with dotted line and the predicted value is represented by a full line.

Table 1 Performance indexes results obtained for the tested models

Model	PI_1	PI_2	Model	PI_1	PI_2
AR(10)	4.12	23.4	ANN-WT [b]	3.9	24.4
AR(10)[a]	6.18	23.9	SVR-LK [c]	4.05	23.1
ANN	4.09	23.3	SVR-GK [d]	3.85	24.2

[a] Recursive least squares with a 0.998 forgetting factor. [b] WT stands for Wavelet Transform. [c] Linear kernel. [d] Gaussian kernel.

4 Conclusion and Further Work

Obtaining good prediction on the long term solar radiation evolution is of major importance in several engineering fields. From solar energy power plants to agriculture there are a myriad of potencial application for good prediction models.

However, due to lack of stationarity and low past (linear) correlation, modelling the solar radiation as a time-series, for long time forecast, is a very difficult problem. Several models and techniques have already been tryout: from simple naive to more elaborated hybrid strategies. From our experience it seems to be some kind of non-linear relationship between the present observation and the past data since the non-linear models used have increased performance when compared to linear ones, although, in long prediction range, all the models reveal severe divergence when targeting with the observed data.

Nevertheless the SVR strategy, even if timid in the performance indexes revealed, was able to give slightly better prediction than the other tested methods. Currently the autors are engaged with an hybrid strategy that involves hidden Markov models in order to bypass the non-stationarity behaviour of time-series.

References

1. Cherkassky, V., Ma, Y.: Practical Selection of SVM parameters and Noise Estimation for SVM Regression. Neural Networks 17, 113–126 (2004)
2. Coelho, J., Cunha, J.B., Oliveira, P.B.: Greenhouse Air Temperature Control using the Particle Swarm Optimisation Algorithm. COMPAG - Computers and Electronics in Agriculture 49(3), 330–344 (2005)
3. Coelho, J., Cunha, J.B., Oliveira, P.B.: Solar radiation prediction using wavelet decomposition. In: 8th Portuguese Conference on Automatic Control - CONTROLO 2008 (2008)
4. Gunn, S.R.: Support vector machines for classification and regression. Technical report. University of Southampton (1998)
5. Kuhn, H.W., Tucker, A.W.: Nonlinear programming. In: Proceedings of 2nd Berkeley Symposium, pp. 481–492. University of California Press, Berkeley (1951)

6. Müller, K.-R., Smola, A., Rätsch, G., Schölkopf, B., Kohlmorgen, J., Vapnik, V.: Predicting Time Series with Support Vector Machines. In: Gerstner, W., Hasler, M., Germond, A., Nicoud, J.-D. (eds.) ICANN 1997. LNCS, vol. 1327, pp. 999–1004. Springer, Heidelberg (1997)
7. Van Straten, G.: On-line optimal control of greenhouse crop cultivation. Acta Hort. 406, 203–212 (1996)
8. Vapnik, V.N.: Statistical Learning Theory. John Wiley and Sons, New York (1995)

Evaluating the Low Quality Measurements in Lighting Control Systems

Jose R. Villar, Enrique de la Cal, Javier Sedano, and Marco García

Abstract. In real world processes in the industry or in business, where the elements involved generate data full of noise and biases, improving the energy efficiency represents one of the main challenges. In other fields as lighting control systems, the emergence of new technologies, such as the Ambient Intelligence, also degrades the quality data introducing linguistic values. In this contribution we propose the use of the novel genetic fuzzy system approach to obtain classifiers and models able to manage low quality data to improve the energy efficiency. The problem is introduced through the experimentation to figure out how significant the improvement of managing the low quality data can be.

1 Introduction

In general, multi-agent architecture and the distribution among the intelligent devices of the control and the optimisation of the decisions may improve the energy efficiency [16], which represents a big challenge in different engineering fields as efficient design and operation [9], modeling and simulation [3], etc.

Jose R. Villar
University of Oviedo, Campus de Viesques s/n 33204 Gijón, Spain
e-mail: villarjose@uniovi.es

Enrique de la Cal
University of Oviedo, Campus de Viesques s/n 33204 Gijón, Spain
e-mail: delacal@uniovi.es

Javier Sedano
University of Oviedo, Avda. Cantabria s/n 09006 Burgos, Spain
e-mail: jsedano@ubu.es

Marco García
University of Oviedo, Campus de Viesques s/n 33204 Gijón, Spain
e-mail: marco@uniovi.es

E. Corchado et al. (Eds.): SOCO 2010, AISC 73, pp. 119–126.
springerlink.com

In what follows, the field of lighting control systems will be analysed for the sake of simplicity, although the main conclusions can be extended to any other area.

The lighting control systems have been studied in depth: the simulation issues [3], sensor processing and data improvement [4], the effect of daylight in the energy efficiency [7], among others. Moreover, the improvement in the energy efficiency and its measurement have been analysed in [7, 9].

In a lighting control system (see Fig. 1), the lighting system controller is the software responsible for co-ordinating the different islands and for integrating the information from the Building Management Systems (BMS). In each island, a controller establishes the operation conditions of all the controlled ballasts according to the sensor measurements and the operation conditions given by the lighting system controller.

Nevertheless, the meta-information in the data gathered from processes is rarely used, and it is mainly related to non-stochastic noise. This meta-information related with the low quality data can also be due to the precision of the sensors and to the emergence of new technologies such as Ambient Intelligence and the user profiles. In our opinion, the use of Genetic Fuzzy Systems (GFS) could improve the issues related with energy sharing and efficiency in distributed systems. We propose using the GFS able to deal with the meta-information to achieve better energy efficiency results.

In this research we show how the uncertainty in real world problems can be observed, specifically, in lighting systems. We propose the use of a novel method for learning GFS with low quality data for improving the energy efficiency in distributed systems taking advantage of the meta-data due to low quality data. The remainder of this manuscript is as follows. Firstly, the uncertainties in real world problems will be shown using the simulation of

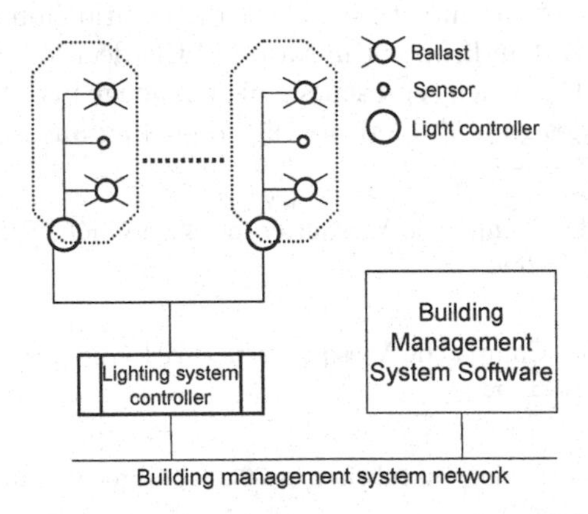

Fig. 1 The schema of a lighting control system. Each island includes a closed loop controller with the controlled gears, the luminosity sensors, and the presence sensors. The lighting system controller is the responsible for co-ordinating the islands.

lighting control systems. Then, a short review of the literature concerned with considering the low quality data in modeling and in designing indexes is shown. Finally, some conclusions in how to manage such low quality data are presented.

2 Low Quality Data in Lighting Systems and the Energy Efficiency

Lighting control systems aim to set the electric power consumption for the ballast in the installation so the luminance complies with the regulations. In such systems, the luminance is measured through light sensors. Variables as the presence of inhabitants are also used in lighting control systems. Even though there are more variables, the relevance of the former is higher as it is used as the feedback in the lighting control loop. Nevertheless, the output of such sensors is highly dependant of the sunlight, the magnitude varies from one sensor to other, the repeatability is a compromise, etc. Consequently, the output of the sensors is usually filtered and then used as the feedback of the control loop, always as a crisp value.

Simulation of lighting systems has been widely studied, mainly to improve the energy efficiency [3, 7]. A lighting system simulation needs to simulate the light measured in a room when a total electric power is applied for lighting. A simulation will use models to estimate the response of the light sensors. The main objective in simulation is to set and tune PID controllers for light control systems. As before, light sensors measurements are considered crisp values, and so are the inputs and the outputs of the light sensor models. To our knowledge, no model has been obtained including the meta-information due to low quality data and, thus, the effect of the daylight and other variables are introduced artificially -i.e., by considering such information within the input data set.

Let us consider one simple case. Let us suppose the simulation of the lighting system shown in Fig. 2, where there is a simple room with one light sensor installed and the light gears accomplishing the regulations. Where to fix the light sensor is of great significance as the shorter the distance from the light sensor to the windows the higher the daylight influence in the light measurements. On the other hand, the daylight should be estimated from the inner light sensors when no daylight sensors are available. Let us suppose the light sensor installed next to the window, with the light sensor being a LDR (light dependant resistor). Let us also consider a lighting control system that allows regulations on the electric power for lighting, so it is possible to introduce several steps, say 0%, 33%, 66% and 100% of the total power installed. Finally, there was a blind that can be opened or closed. In this scenario, several experiments were carried out. In all of them, the controlled variable was the percentage of electric power for lighting and the output from the light sensor (as a voltage value) was sampled.

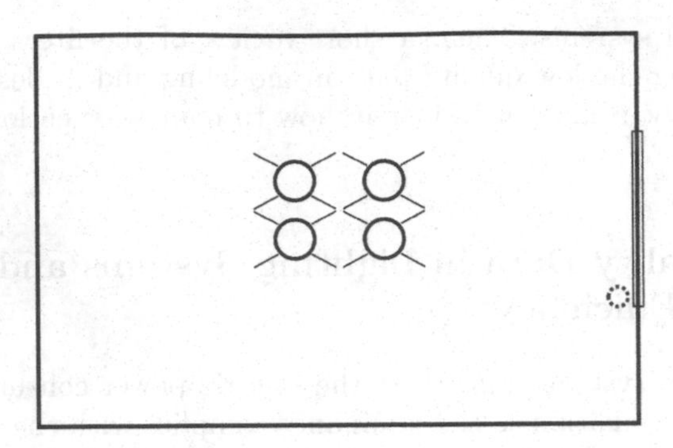

Fig. 2 The lighting system to simulate. Different places for the light sensor are proposed. The light measured will differ from one case to another.

The first experiment was the step response increasing and decreasing the controlled variable with the blind closed, measuring the light sensor output. This experiment was carried twice for each sensor, and repeated for five different sensors. In Fig. 3 the results are presented. As can be seen, the measurements are highly dependant of the sensor itself, but also the hysteresis behaviour can be perceived. In all the figures normalised values are presented for the sake of clearness.

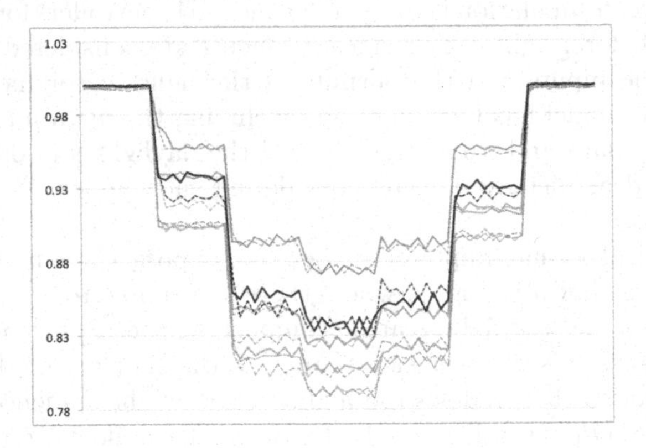

Fig. 3 The normalised output of the different light sensors

Anyway, this is a measure linear with the resistance of the LDR, but what about the luminance? To calculate the luminance $R = C_0 \cdot L^{-\gamma}$ applies, where R is the measured resistance, C_0 and γ are constants for each LDR. LDR's are characterised with a range of possible values of resistance for a luminance of 10 lux (R_{10}). In case of the ones used in the experiments, R_{10} varies from $[8, 20]k\Omega$ and γ is 0.85, typically. Thus, given a sensor, a minimum and a maximum of the luminance can be obtained if a resistance value is measure (see Fig. 4. A measurement of resistance corresponds with an interval of luminance. I.e., if $r = 0.6$ induces a luminance in the range $[\sim 0.04, \sim 0.15]\%$, where the hysteresis does not allow to establish the exact values of luminance. And for the up/down step responses there is a wide margin of possible values of the luminance as shown in Fig. 5.

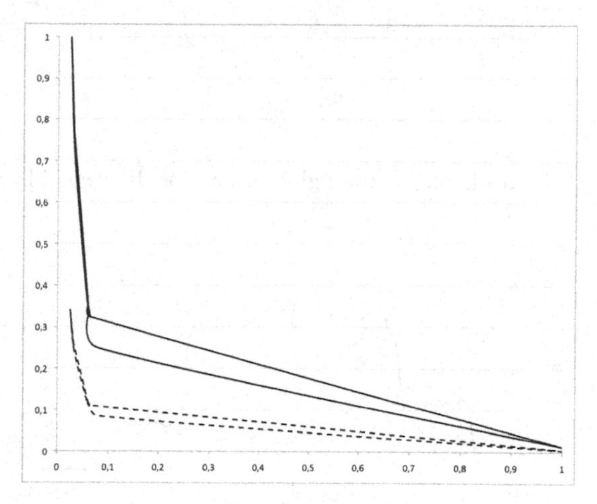

Fig. 4 The resistance against the luminance for the minimum and maximum bounds, both variables in percentage of their maximum value

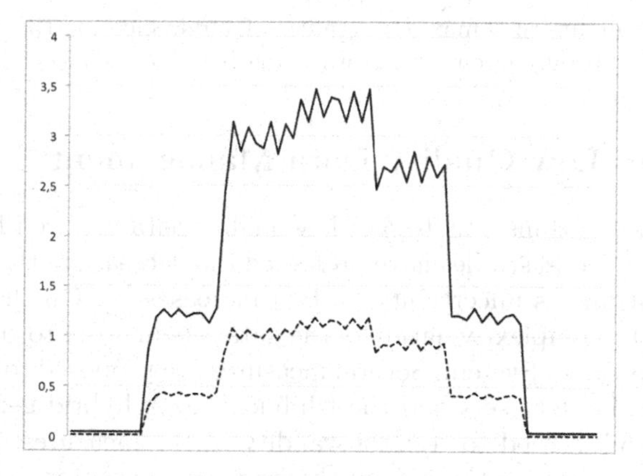

Fig. 5 The minimum and maximum values of luminance for the step responses

The second experiment includes the opening of the blind. Three positions were considered: totally closed, half open and totally open. With the same characteristics as in the first experiment, the step responses for the up/down sequence for one of the sensors is shown in Fig. 6 and Fig. 7.

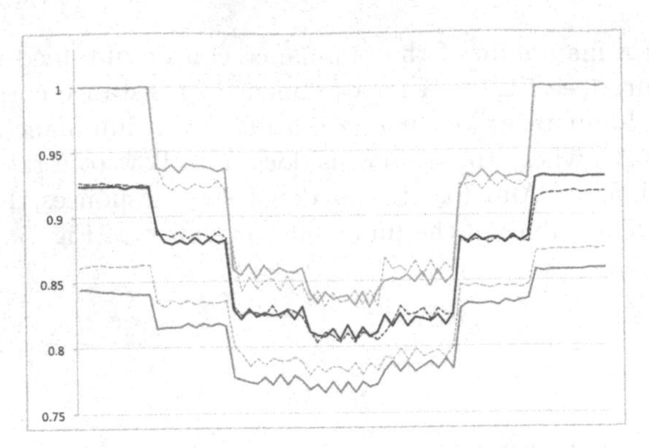

Fig. 6 The normalised output of the light sensors for different blind positions

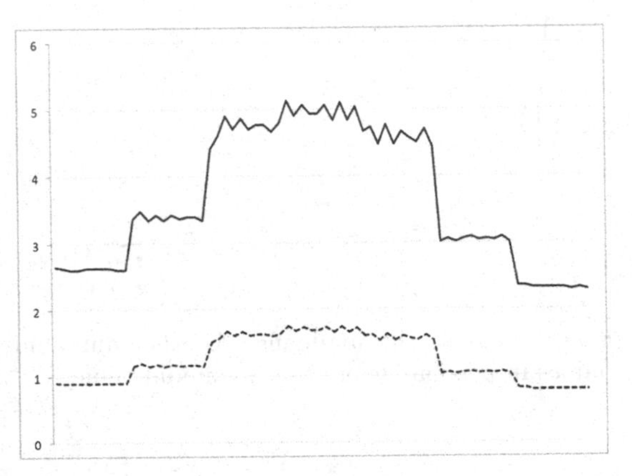

Fig. 7 The minimum and maximum values of luminance for the step responses when the blind is totally open for a stormy weather day

3 Issues in Low Quality Data Management

The need for algorithms able to face low quality data is a well-known fact in the literature. Several studies have presented the decrease in the performance of crisp algorithms as uncertainty in data increases [5]. On the other hand, [10] analyses the complexity nature of the data sets in order to choose the better Fuzzy Rule Based System. Several measures are proposed to deal with the complexity of the data sets and the Ishibuchi fuzzy hybrid genetic machine learning method is used to test the validity of the measures. This research also concludes in the need to extend the proposed measures to deal with low quality data. With low quality data we refer to the data sampled in presence of non-stochastic noise or obtained with imprecise sensors. It is worth noting that all the sensors and industrial instrumentation can be regarded as low quality data. In our opinion, one of the most successful researches in soft computing dealing with low quality data is detailed in [2, 12]. In these works the mathematical basis for designing vague data awareness genetic fuzzy systems -both classifiers and models- is shown. The low quality data are assumed as fuzzy data, where each $\alpha-$cut represents an interval value for each data.

Finally, it is worth pointing out that the fitness functions to train classifiers and models are also fuzzy valued functions when faced with low quality data. Hence the learning algorithms should be adapted to such fitness functions [14]. The guidelines for learning regression models based on low quality data is detailed in [13]. As it has been shown in previous Section, there is an evidence that light controllers can not be optimum, thus the energy efficiency in lighting control systems can be improved if such low quality data is considered. Specifically, the use of the methodology using classical control theory proposed by different authors as [3] may not be the best choose. In our opinion, the use of GFS able to manage low quality data in obtaining models for simulation of lighting systems would help in the integration of the meta-information. The use of GFS allows determining behaviour laws and interpretability of the phenomena. Moreover, if low quality data is included in obtaining the models of the sensors the controllers would be robust to such variability.

4 Conclusions

Improving the energy efficiency represents a challenge in the real world applications, especially distributed systems within building management systems. Meta-information as low quality data is included in all the data gathered from such processes. Meta-information refers to the information that is present in a process but rarely considered such as the data non-stochastic noise or the sensor precision and calibration, but also the ambiguity in the linguistic and crisp data. In this work it is proposed the use of the extended GFS to manage the low quality data in the energy efficiency improvement. To illustrate the idea, the uncertainty in the luminance measurements from light sensors is analysed. We expect that fuzzy data awareness GFS will outperform the modelling and simulation process, so better controllers can be obtained.

Acknowledgements. This research work is been funded by Gonzalez Soriano, S.A. by means of the the CN-08-028-IE07-60 FICYT research project and by Spanish M. of Science and Technology, under the grant TIN2008-06681-C06-04.

References

1. Bernal-Agustín, J.L., Dufo-López, R.: Techno-economical optimization of the production of hydrogen from PV-Wind systems connected to the electrical grid. Renewable Energy 35(4), 747–758 (2010)
2. Couso, I., Sánchez, L.: Higher order models for fuzzy random variables. Fuzzy Sets and Systems 159, 237–258 (2008)

3. de Keyser, R., Ionescu, C.: Modelling and simulation of a lighting control system. Simulation Modelling Practice and Theory (2009), doi: 10.1016/j.simpat.2009.10.003
4. Doulos, L., Tsangrassoulis, A., Topalis, F.V.: The role of spectral response of photosensors in daylight responsive systems. Energy and Buildings 40(4), 588–599 (2008)
5. Folleco, A.A., Khoshgoftaar, T.M., Van Hulse, J., Napolitano, A.: Identifying Learners Robust to Low Quality Data. Informatica 33, 245–259 (2009)
6. Gligor, A., Grif, H., Oltean, S.: Considerations on an Intelligent Buildings Management System for an Optimized Energy Consumption. In: Proceedings of the IEEE Conference on Automation, Quality and Testing, Robotics (2006)
7. Hviid, C.A., Nielsen, T.R., Svendsen, S.: Simple tool to evaluate the impact of daylight on building energy consumption. Solar Energy (2009), doi:10.1016/j.solener.2008.03.001
8. Houwing, M., Ajah, A.N., Heijnen, P.W., Bouwmans, I., Herder, P.M.: Uncertainties in the design and operation of distributed energy resources: The case of micro-CHP systems. Energy 33(10), 1518–1536 (2008)
9. Li, D.H.W., Cheung, K.L., Wong, S.L., Lam, T.N.T.: An analysis of energy-efficient light fittings and lighting controls. Applied Energy 87(2), 558–567 (2010)
10. Luengo, J., Herrera, F.: Domains of competence of fuzzy rule based classification systems with data complexity measures: A case of study using a fuzzy hybrid genetic based machine learning method. Fuzzy Sets and Systems 161, 3–19 (2010)
11. Qiao, B., Liu, K., Guy, C.: A Multi-Agent System for Building Control. In: IAT 2006: Proceedings of the IEEE/WIC/ACM international conference on Intelligent Agent Technology, pp. 653–659. IEEE Computer Society, Los Alamitos (2006)
12. Sánchez, L., Couso, I.: Advocating the Use of Imprecisely Observed Data in Genetic Fuzzy Systems. IEEE Transactions on Fuzzy Systems 15(4), 551–562 (2007)
13. Sánchez, L., Otero, J.: Learning Fuzzy Linguistic Models from Low Quality Data by Genetic Algorithms. In: Proceedings of the IEEE Internacional Conference on Fuzzy Systems FUZZ-IEEE 2007 (2007)
14. Sánchez, L., Couso, I., Casillas, J.: Genetic Learning of Fuzzy Rules based on Low Quality Data. Fuzzy Sets and Systems (2009)
15. Villar, J.R., Pérez, R., de la Cal, E., Sedano, J.: Efficiency in Electrical Heating Systems: An MAS real World Application. In: Proceedings of the 7th International Conference on Practical Applications of Agents and Multi-Agent Systems (PAAMS 2009). LNCS, vol. 55, pp. 460–469 (2009)
16. Villar, J.R., de la Cal, E., Sedano, J.: A fuzzy logic based efficient energy saving approach for domestic heating systems. Integrated Computer-Aided Engineering 16(2), 151–164 (2007)

Soft Computing Models for an Environmental Application

Ángel Arroyo, Emilio Corchado, and Verónica Tricio

Abstract. In this interdisciplinary research several statistical and soft computing models are applied to analyze a case study related to inmissions of atmospheric pollution in urban areas. The research analyzes the impact on atmospheric pollution of an extended bank holiday weekend in Spain and the way in which meteorological conditions affect pollution levels. After classifying atmospheric pollution levels in relation to the days of the week, we analyze the way in which these may be influenced by atmospheric conditions. The case study is based on data collected by a station at the city of Burgos, which forms part of the pollution measurement station network within the Spanish Autonomous Region of Castile-Leon.

Keywords: Artificial neural networks, soft computing, meteorology, atmospheric pollution, statistical models.

1 Introduction

It has been accepted for some years now that air pollution not only represents a health risk, but that it also, for example, reduces food production and vegetative growth due to its negative effects on photosynthesis.

Ángel Arroyo
Department of Civil Engineering, University of Burgos,
c\Francisco de Vitoria s/n 09006 Burgos, Spain
e-mail: aarroyop@ubu.es

Emilio Corchado
Department of Computer Science and Automatic, University of Salamanca,
Plaza de La Merced s/n, 37008, Salamanca, Spain
e-mail: escorchado@usal.es

Verónica Tricio
Department of Physics, University of Burgos, Burgos, Spain
e-mail: vtricio@ubu.es

E. Corchado et al. (Eds.): SOCO 2010, AISC 73, pp. 127–135.
springerlink.com

Systematic measurements in Spain, which are usually taken within large cities, are fundamental, due to the health risks caused by high levels of atmospheric pollution. European legislation, will in the long term establish how and where such pollutants should be measured.

Soft computing [1, 2] consists of various technologies which are used to solve inexact and complex problems. It is used to investigate, simulate, and analyze complex issues and phenomena in an attempt to solve real-world problems.

2 Soft Computing and Statistical Models

Several statistical and artificial neural networks (ANNs) have been applied and compared in this study to identify optimal performance within the *meteorology and pollution case study presented in this study.*

Principal Components Analysis (PCA). PCA [3] gives the best linear compression of the data in terms of least mean square error and can be implemented by several artificial neural networks [4, 5]. The basic PCA network applied in this study is described by (Eq. 1) and (Eq. 2): an N-dimensional input vector at time t, $x(t)$, and an M-dimensional output vector, y, with W_{ij} being the weight linking input j to output i, and η being the learning rate. Its activation and learning may be described as follows:

Feedforward step, (Eq. 1):

$$y_i = \sum_{j=1}^{N} W_{ij} x_j \quad , \forall\ i \tag{1}$$

Feedback step, (Eq. 2):

$$e_j = x_j - \sum_{i=1}^{M} W_{ij} y_i \tag{2}$$

Change weights, (Eq. 3):

$$\Delta W_{ij} = \eta e_j y_i \tag{3}$$

This algorithm is equivalent to Oja's Subspace Algorithm [6], (Eq. (4)):

$$\Delta W_{ij} = \eta e_j y_i = \eta (x_j - \sum_{k} W_{kj} y_k) y_i \tag{4}$$

Maximum Likelihood Hebbian Learning (MLHL) [7] identifies interesting directions or dimensions by maximising the probability of the residuals under specific probability density functions that are non-Gaussian. Considering an N-dimensional input vector (X), and an M-dimensional output vector (y), with W_{ij} being the weight (linking input j to output i), then MLHL can be expressed as:

1 Feed-forward step (Eq. 5):

$$y_i = \sum_{j=1}^{N} W_{ij} x_j, \forall i \cdot \tag{5}$$

2 Feedback step (Eq. 6):

$$e_j = x_j - \sum_{i=1}^{M} W_{ij} y_i, \forall j \tag{6}$$

3 Weight change (Eq. 7):

$$\Delta W_{ij} = \eta . y_i . sign\left(e_j\right) | e_j |^{p-1} . \tag{7}$$

Where: η is the learning rate, b the bias parameter, and p a parameter related to the energy function.

Cooperative Maximum Likelihood Hebbian Learning (CMLHL) [8, 9] is an extended version of MLHL adding lateral connections which have been derived from the Rectified Gaussian Distribution [10].

Consider an N-dimensional input vector (x), an M-dimensional output vector (y) and a weight matrix W, where the element W_{ij} represents the relationship between input x_j and output y_i, then as is shown in [9, 11], CMLHL can be performed as a four-step procedure:

Feed-forward step, outputs are calculated (Eq. 8):

$$y_i = \sum_{j=1}^{N} W_{ij} x_j, \forall i \tag{8}$$

Lateral activation passing step, (Eq. 9):

$$y_i(t+1) = \left[y_i(t) + \tau(b - Ay)\right]^+ \tag{9}$$

Feedback step, (Eq. 10):

$$e_j = x_j - \sum_{i=1}^{M} W_{ij} y_i, \forall j \tag{10}$$

Weights update step, learn the neural network, (Eq. 11):

$$\Delta W_{ij} = \eta . y_i . sign\left(e_j\right) | e_j |^{p-1} \tag{11}$$

Where t is the temperature, $[\ ]^+$ is necessary to ensure that the y-values remain in the positive quadrant, η is the learning rate, τ is the "strength" of the lateral connections, b the bias parameter, p a parameter related to the energy function, and A is a symmetric matrix used to modify the response to the data.

3 A Meteorological Case Study

A meteorological case study is presented here that worked with data on the evolution of different pollution and meteorological parameters using the records of an air quality control station. The aforementioned station [12] is situated in the urban area of the city of Burgos. After analyzing all the weeks in 2007, a week with a national bank holiday weekend in December was selected because many people in Spain go on holiday and travel over this period, greatly decreasing traffic flows in large cities and almost completely curtailing industrial activity.

In this research, the following variables were analyzed: SO_2, NO, NO_2, PM10 as pollutant parameters; and wind direction, wind speed, dry temperature, relative humidity, atmospheric pressure and solar radiation as meteorological parameters.

This study examines the performance of several statistical and soft computing methods when analyzing the above-mentioned pollution variables, in order to track the evolution of air pollution over a significant period of time and the influence of atmospheric conditions in air quality.

Thus, as a first step, the main aim is to identify the extent to which air quality is affected on days with low industrial activity and reduced traffic flows. In a second step the evolution of meteorological conditions throughout the week will be analyzed. Finally, the influence of meteorological conditions on air pollution over the same period of time will also be studied.

4 Experiments and Results

The study, which forms part of a more ambitious project [13, 14], is based on meteorological and pollution data sets recorded at fifteen-minute intervals: a daily total of 96 records for all data in 2007. On this occasion, for presentation purpose, hourly averages are taken. There are 168 samples per week.

The information represented at each point is visually labeled from Fig. 1 to Fig. 3, which shows the time in 24h format and a weekday initial (e.g., 6M means 6 am - Monday). All data was normalized for the study.

In this first step the evolution of air pollution throughout the week is presented and only pollutant variables are analyzed.

Fig. 1 (a). PCA. This method identifies two main clusters (C_1 and C_2). Cluster C_2 groups most of the weekday samples and it is difficult to analyze its structure. Cluster C_1 is a group of scattered points that correspond in this case to the samples for Monday (M), which show the highest pollution levels, throughout the entire week.

Fig. 1 (b) shows the projection generated by the MLHL method. Cluster C_1 represents the samples showing the highest pollution on Monday (M) evening, in the same way as was identified by PCA (Fig. 1 (a)). Cluster C_2 (Fig. 1 (a)) corresponds to clusters C_{2a} and C_{2b} in (Fig. 1 (b)). The samples for cluster C_{2a} have lower pollution levels than the samples for cluster C_1. These samples correspond to Tuesday (T) and Wednesday (W). In cluster C_{2b}, the points represent lower levels of pollution, corresponding to Thursday (Th), Friday (F), Saturday (S) and Sunday (Su). Thursday (Th) and Friday (F) were the days associated with the two national bank holidays: Constitution Day and the Day of the Immaculate Virgin Pilar, respectively. On these days, industrial activity is greatly curtailed and many people traditionally travel long distances.

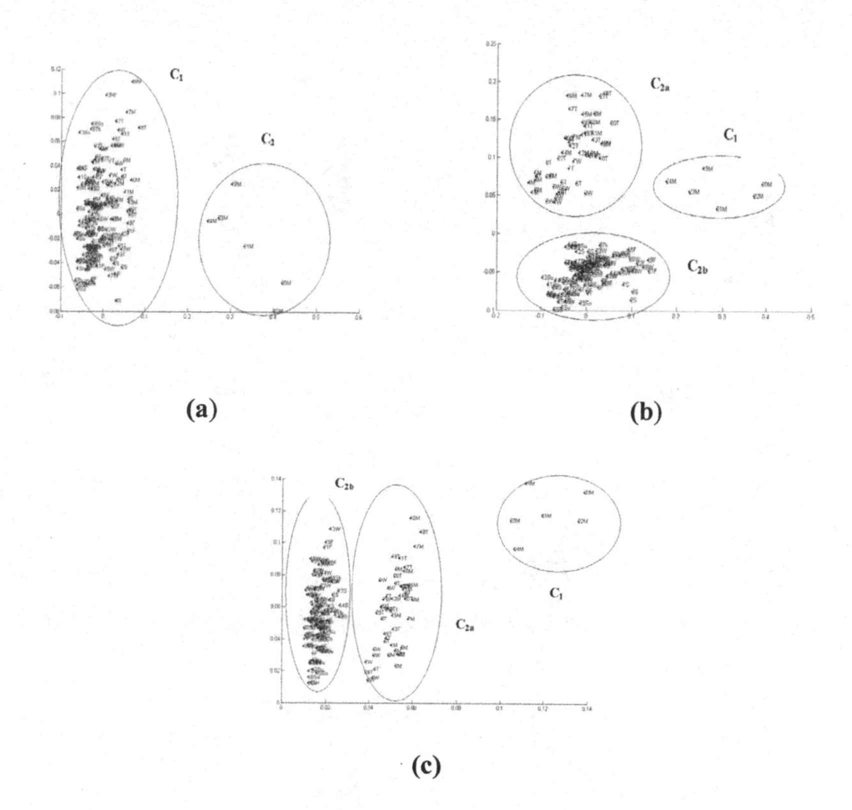

Fig. 1 (a) PCA projection, (b) MLHL projection and (c) CMLHL projection of hourly pollution parameters

Fig. 1 (c) shows the projection generated by the CMLHL method. The difference with MLHL, (Fig 1 (a)), is that CMLH is able to identify a new sub-cluster of samples. Cluster C_{2a} in (Fig. 1 (a)) corresponds to clusters C_{2a1} and C_{2a2} in (Fig. 1 (c)). Samples in C_{2a1} have higher levels of pollution than samples in C_{2a2}.

In the second step, the evolution of meteorological conditions throughout the week is presented and the meteorological variables are analyzed.

Fig. 2 (a) shows the PCA projection. Cluster C_1 is formed with a few records which correspond to the midday of Sunday (Su). In these hours, solar radiation reached the highest values in the whole week. In cluster C_2 the samples out of the main cloud correspond to periods in the afternoon of Saturday (Sa) and Sunday (Su) and to the morning and midday of Monday (M), Tuesday (T), Wednesday (W) and Thursday (Th), in which solar radiation was also quite high.

Fig. 2 (b) shows the projection obtained by MLHL method. This soft computing model identifies the C_1 cluster which contains the same records as the equivalent PCA clusters (Fig 2 (a)), and clusters C_{2a} and C_{2b} which corresponds to

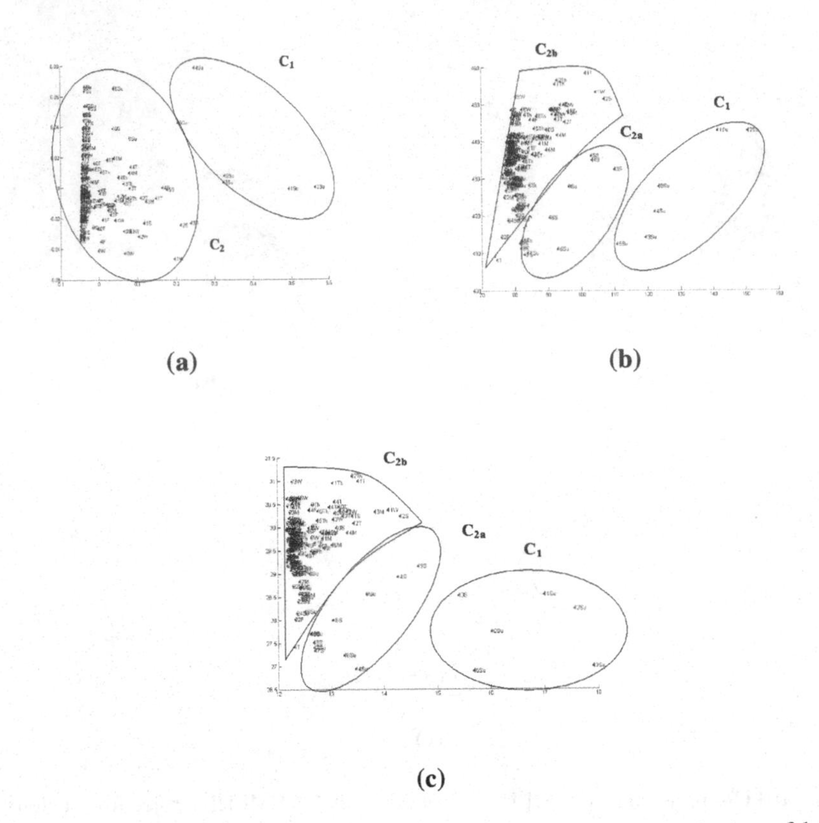

Fig. 2 (a) PCA projection, (b) MLHL projection and (c) CMLHL projection of hourly atmospheric parameters

the C_2 clustering (Fig 2 (a)). Cluster C_{2a} corresponds to periods during Saturday (S) and Sunday (Su) afternoon where solar radiation records high values.

Fig. 2 (c) shows the projection generated by the CMLHL method. The results obtained by applying CMLH are really similar to those generated by MLHL, (Fig 2(b)). The only difference this time is that the records in cluster C_{2b} are sparser than those in cluster C_{2b} (Fig 2(b)).

Finally, the influence of meteorological conditions on air pollution in the same period of time is analyzed. This time, all the variables both, atmospheric and pollutant parameters are combined.

PCA (Fig. 3 (a)) identifies three main clusters. Cluster C_1 is formed by a few records which correspond to midday on Sunday (Su), at which time, the highest solar radiation values in the week were registered. C_2 is a group of scattered points that correspond to the samples taken on Monday (M), which have the highest pollution throughout the entire week. C_3 contains the rest of the records.

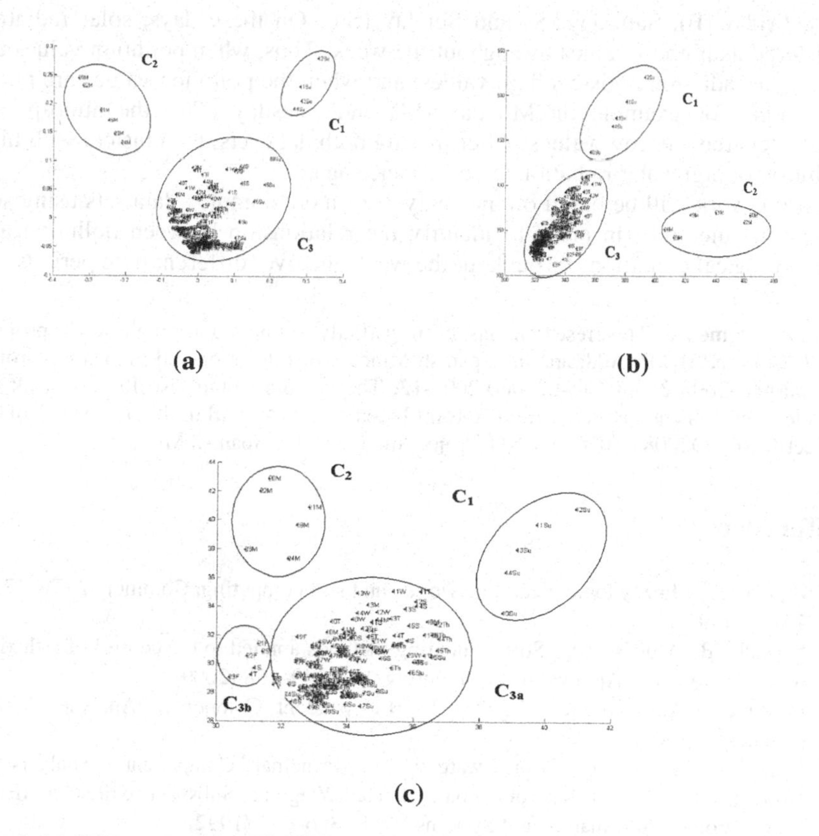

Fig. 3 (a) PCA projection, (b) MLHL projection and (c) CMLHL projection of hourly atmospheric and pollutant parameters

MLHL (Fig. 3 (b)) identifies the same three clusters identified by PCA, (Fig. 3 (a)).

CMLHL (Fig. 3 (c)) yields a set of results which are better than those yielded by MLHL, (Fig 3(b)). This time the samples contained in the cluster C_3 (Fig 3(a) and Fig 3 (b)) are split into clusters C_{3a} and C_{3b}. Cluster C_{3b} corresponds to a time on Friday (F) night with high temperature values.

5 Conclusions and Future Works

The PCA model provides an initial approximation to the internal structure of the data, but does not offer the best information on hourly pollution and atmospheric readings. MLHL and CMLHL offer a better response and both show very similar results. Comparing the three methods, CMLHL generated the most accurate response. The cluster with the lowest levels of pollution has been identified in this study and, as expected, it corresponds to the national bank holiday days: Thursday

(Th), Friday (F), Saturday (S) and Sunday (Su). On these days, solar radiation registered its highest values throughout the week. Thus, when pollution values are low, solar radiation registers high values, and when the pollution values are really very high, for example on Monday (M) and Tuesday (Tu), the atmospheric parameters register low values. After mixing both data sets, the clusters with high pollution or high solar radiation appeared once again.

Future work will be based on the analysis of more complex data sets using soft computing methods, in order to identify the relationship between pollution and meteorological conditions throughout the week and over different time periods.

Acknowledgments. This research has been partially supported through JCyL projects BU006A08 and BU035A08, and the Spanish Ministry of Education and Innovation project CIT-020000-2008-2 and CIT-020000-2009-12. The authors would also like to thank the vehicle interior manufacturer, Grupo Antolin Ingenieria, S.A., within the framework of the project MAGNO2008 - 1028.- CENIT Project funded by the Spanish Ministry.

References

1. Zadeh, L.A.: Fuzzy logic, neural networks, and soft computing. Commun. ACM 37(3), 77–84 (1994)
2. Subudhi, B., Morris, A.S.: Soft computing methods applied to the control of a flexible robot manipulator. Applied Soft Computing 9(1), 149–158 (2009)
3. Hyvärinen, A., Karhunen, J., Oja, E.: Independent Component Analysis. Wiley, Chichester (2002)
4. Oja, E., Ogawa, H., Wangviwattana, J.: Principal Components Analysis by Homogeneous Neural Networks, part 1. The Weighted Subspace Criterion, IEICE Transaction on Information and Systems E75D, 366–375 (1992)
5. Fyfe, C., Baddeley, R.: Non-linear data structure extraction using simple Hebbian networks. Biological Cybernetics 72(6), 533–541 (1995)
6. Oja, E.: Neural Networks, Principal Components and Subspaces. International Journal of Neural Systems 1, 61–68 (1989)
7. Fyfe, C., Corchado, E.: Maximum Likelihood Hebbian Rules. In: Proc. of the 10th European Symposium on Artificial Neural Networks (ESANN 2002), pp. 143–148 (2002)
8. Harman, H.: Modern Factor Analysis, 2nd edn. University of Chicago Press, Chicago (1967)
9. Corchado, E., MacDonald, D., Fyfe, C.: Maximum and Minimum Likelihood Hebbian Learning for Exploratory Projection Pursuit. Data Min. Knowl. Discov. 8(3), 203–225 (2004)
10. Seung, S., Socci, N.D., Lee, D.: The Rectified Gaussian Distribution. Advances in Neural Information. Processing Systems 10, 350–356 (1998)
11. Fyfe, C., Corchado, E.: Maximum Likelihood Hebbian Rules. In: Proc. of the 10th European Symposium on Artificial Neural Networks (ESANN 2002), pp. 143–148 (2002)

12. Tricio, V., Viloria, R., Minguito, A: Ozone Measurements. In: Urban and Semi-Rural Sites At Burgos (Spain). Geophysical Research Abstracts, Spain). EGS-AGU-EUG Joint Assembly, vol. 5 (2003); EAE03-A-14249. ISSN: 1029-7006
13. Arroyo, A., Corchado, E., Tricio, V.: Atmospheric Pollution Analysis by Unsupervised Learning. In: Corchado, E., Yin, H. (eds.) IDEAL 2009. LNCS, vol. 5788, pp. 767–772. Springer, Heidelberg (2009)
14. Arroyo, A., Corchado, E., Tricio, V.: Computational Methods for Immision Analysis of Urban Atmospheric Pollution. In: 9th International Conference Computational and Mathematical Methods in science and engineering, Gijón (2009)

GRASP Algorithm for Optimization of Grids for Multiple Classifier System

Tomasz Kacprzak, Krzysztof Walkowiak, and Michał Woźniak

Abstract. In recent years the volume of data used in scientific researches and industry has increased significantly. Distributed computing systems including Grids use the public Internet to share computational resources of research institutions around the world in order to process the data. Due to large data volumes being transferred, network aspects of Grids have become important. In this work we introduce a model of an overlay Grid system, which could be used by the distributed recognition system based on the idea of combining classifiers. We formulate an Integer Programming optimization problem with the objective to minimize the overall cost including processing and data transfer. Next, an effective heuristic algorithm is developed to solve the problem. Results of numerical experiments showing the comparison of the heuristic against solutions provided by CPLEX solver are presented.

1 Introduction

Progress of computer science caused that many institutions collected huge amount of data, which analysis is impossible by human being, e.g. projects devoted High Energy Physics, Bioinformatics, Astronomy and Earth Observation, Climate Modeling etc.. Most of enterprises cannot take advantage of useful information hidden in their enormous databases, because traditional data analysis tools are practicable only on small databases. An useful task is to build a model of data so as to produce classifier for prediction. Designing classification methods for very large databases is currently the focus of intense research [5, 6]. The aim of the classification task, known also as pattern recognition, is to classify the object to one of the predefined categories, on the basis of its feature values. Aforementioned methods are usually applied to the many practical areas like credit approval, prediction of customer behavior, fraud detection, designing IPS/IDS,

Tomasz Kacprzak · Krzysztof Walkowiak · Michał Woźniak
Wrocław University of Technology, Wybrzeże Wyspiańskiego 27,
50-370 Wrocław, Poland
e-mail: {Krzysztof.Walkowiak,Michal.Wozniak}@pwr.wroc.pl

E. Corchado et al. (Eds.): SOCO 2010, AISC 73, pp. 137–144.
springerlink.com

medical diagnosis, to name only a few. Numerous approaches have been proposed to construct efficient classifiers like neural networks, statistical learning, and symbolic learning [2].

One of the most promising approach is Multiple Classifier Systems (MCSs) which concern methods for improving classification accuracy or response time through the use of partitions of original data set for the simple classifiers. We consider such a system as the set of *V* autonomous classifiers which are trained on the basis of mentioned above separated partitions of huge, original database. For each partition of data sets decision about object classification is made independently and the final decision is made on the basis of aforementioned *V* decisions using a fusion method [13]. This approach could produce classifier which accuracy is higher than an individual one based on the whole learning set. The proposition is very flexible and it could be used for the case that database is partitioned from privacy reasons or to speed up the decision making process. One of the most promising approach which could be use for MCSs are Grid networks, which are used to share computational resources required to realize these projects. The increase of data volume in scientific research processing resulted in creation of the term Data Grid [8, 14] – a Grid network, which receives, processes and sends large volumes of data. Examples of such networks are World Large Hadrons Collider Computing Grid and Biomedical Informatics Research Network. The volume of network traffic generated by these networks is becoming significant [15]. For example, average data stream generated by LHC measuring devices is about 500 MB/s [16], and then it is being sent do the Grid for processing. Cost of exploitation of these networks is becoming important. Scientific projects, which use the Grid networks, last for long time and data is generated constantly by measuring devices, telescopes etc. These projects are assigned to the nodes of the network and reserve resources for a long time. Scheduling is centralized and static. Nodes are permanently available online, connected with high bandwidth network links to the public Internet, so it can be said that Grid networks works as overlay network on the top of underlying Internet. According to [18] parameters of access links are in most cases the only network bandwidth limitations.

Most of previous works on the subject of grid scheduling and resource management do not consider comprehensively the networking aspects - usually the unicast transmission is applied and very few constraint are considered [15] and cost of networking is not taken into account. For that reason in this work we focus on scheduling of computational tasks according to constraints following from network (access links capacity) and Grid (processing power). The objective is to minimize cost of the system compromising transfer cost and processing cost. The Integer Programming model of the overlay Grid network is formulated. Comparing to our previous paper on the same topic [17], in this work we introduce a new Integer Programming formulation of the problem and propose an effective heuristic algorithm based on the GRASP approach. Moreover, numerical experiments are used to verify the effectiveness of the heuristic against the optimal results.

The paper is organized as follows. In section 2 we present idea of multiple classifier systems. Section 3 formulates overlay grid cost minimization problem. In the next section we propose an heuristic algorithm to solve these optimization problem. Section 5 contains results of computer experiments and the last section concludes this work.

2 Models of MCSs

In many review articles MCSs have been mentioned as one of the most promising in the field of pattern recognition [11]. In this conceptual approach, the main effort is concentrated on combining knowledge of the set of elementary classifiers.

There is a number of important issues while building the aforementioned multiple classifier systems. Firstly, how should classifiers be selected. Combining similar classifiers should not contribute much to the system being constructed, apart from increasing the number of computations. So it seems interesting to select members of a committee with possibly different components.

Another important issue is the choice of a collective decision making method. The first group of methods includes algorithms for classifier fusion at the level of their responses [12]. Initially only majority voting schemes were implemented, but in later works more advanced methods were proposed like voting based on weighting the importance of decisions coming from particular committee members [13]. The second group of collective decision making methods exploit classifier fusion based on discriminant analysis, the main form of which are the posterior probability estimators, associated with probabilistic models of a given pattern recognition task [1, 3].

The great advantage of MCSs is that they could be implemented in distributed computing environment, e.g. each individual classifier could be started in the separated node of computing networks.

3 Model of Overlay Grid System

The Grid system consists of clusters (computing systems) – represented as nodes denoted using the subscript index $v = 1,2,\ldots,V$. The computing systems works on the top of an overlay network, i.e. each node (cluster) is connected to the overlay network by an access link with download capacity d_v and upload capacity u_v, expressed in bps. The maximum processing rate of node v, i.e. the number of computational tasks that node v can calculate in one second is denoted as p_v. The processing cost of one computational uniform task in node v is denoted by ψ_v. The transfer cost between nodes w and v is denoted by ζ_{wv}. Each computational project (associated with one pattern recognition task based on the pool of given classifiers) $r = 1,2,\ldots,R$ to be computed in the Grid is described by the following parameters. The number of uniform computational tasks (simple classifier) in project r is n_r, what means that decision is made on the basis of n_r simple classifiers. The input data is transmitted with rate a_r from the source node to one or more computing nodes. The output data is sent with rate b_r from the computing nodes to one or more destination nodes. We make an assumption that the computational project is established for a long time (days, months). Thus, the input and the output data associated with the project is continuously generated and transmitted. We formulate the problem as Integer Program.

Overlay Grid Cost Minimization (OGCM) Problem
indices

$r = 1,2,\ldots,R$ projects
$v,w,z = 1,2,\ldots,V$ nodes (clusters of the grid)

constants

d_v download capacity of node v (b/s)
u_v upload capacity of node v (b/s)
ψ_v processing cost of one computational task in node v
p_v maximum processing rate of node v
ζ_{wv} transfer cost of 1 b/s from node w to node v
n_r number of tasks in project r
a_r transmit rate of input data per one task in project r (b/s)
b_r transmit rate of output data per one task in project r (b/s)
s_{vr} = 1 if node v is the source of project r; 0 otherwise
t_{vr} =1 if node v is the destination of project r; 0 otherwise

variables

x_{rwv} the number tasks of project r that are transmitted from source node w to computing node v (integer)
y_{rwv} the number of output tasks of project r that are transmitted from computing node w to destination node v (integer)

objective

$$\min \quad F = \textstyle\sum_r\sum_w\sum_v x_{rwv}\psi_v + \sum_r\sum_w\sum_v(a_r x_{rwv} + b_r y_{rwv})\zeta_{wv} \tag{1}$$

subject to

$$\textstyle\sum_r\sum_w x_{rwv} \le p_v \quad v = 1,2,\ldots,V \tag{2}$$

$$\textstyle\sum_r\sum_w a_r x_{rwv} + \sum_r\sum_w b_r y_{rwv} \le d_v \quad v = 1,2,\ldots,V \tag{3}$$

$$\textstyle\sum_r\sum_v a_r x_{rwv} + \sum_r\sum_v b_r y_{rwv} \le u_w \quad w = 1,2,\ldots,V \tag{4}$$

$$\textstyle\sum_v x_{rwv} = s_{rw} n_r \quad r = 1,2,\ldots,R \quad w = 1,2,\ldots,V \tag{5}$$

$$\textstyle\sum_w y_{rwv} = t_{rv} n_r \quad r = 1,2,\ldots,R \quad v = 1,2,\ldots,V \tag{6}$$

$$\textstyle\sum_w x_{rwv} \ge y_{rvz} \quad r = 1,2,\ldots,R \quad v = 1,2,\ldots,V \quad z = 1,2,\ldots,V \tag{7}$$

The objective (1) is the cost of the system compromising the computing (processing) cost and the transfer cost. Since each node has a limited processing speed (power) dedicated to computations of the considered job, we add the constraint (2), which guarantees that each node cannot be assigned with more tasks to calculate that it can process. (3) and (4) denote the download and upload capacity constraint, respectively. Condition (5) assures that for each project $r = 1,2,\ldots,R$ each task of project r is assigned to exactly one node v. To meet the requirement that each destination node of project r receives the output data (results of computations) we add constraint (6). Finally, constraint (7) assures that output data can be only uploaded by computation nodes. For more information on modeling of overlay grids refer to [17].

4 Heuristic Algorithm

In this section we propose a new heuristic algorithm solving the OGCM problem formulated as (1)-(7). The algorithm is a based on GRASP (Greedy Randomized Adaptive Search Procedure) approach [4, 7, 10]. It uses a particular representation of the solution based on allocations. The Mixed Objective Function (MOF), comprising of cost and resource consumption assessment, is used to evaluate allocations. Finally, the algorithm consists of two modules: first generates randomized greedy solutions, and second is responsible for parameter adaptation.

In the beginning we introduce a new notation of the OGCM problem which is necessary to present the algorithm. Let $P(V,R,\zeta)$ denote the problem given by (1)-(7). We assume that ω_{vr} denotes the allocation of one computational task of project r to node v. Solution $S(P)$ of problem P is a set of allocations ω_{vr}. The single allocation ω_{vr} can occur in solution $S(P)$ more than once what means that more than one task of project r can be assigned to node v. Let $\omega_{vr}(S(P))$ denote the number of single allocations ω_{vr} in solution $S(P)$. Thus variables x_{rwv} and y_{rwv} can be derived from allocations in the following way $x_{rwv} = s_{vr}\omega_{vr}(S(P))$ and $y_{rwv} = t_{wr}\omega_{vr}(S(P))$.

The cost of allocation ω_{vr} is given by the following formula

$$f(\omega_{vr}) = \psi_v + \textstyle\sum_w s_{wr}a_r\zeta_{wv} + \sum_w t_{wr}b_r\zeta_{vw} \qquad (8)$$

The first element refers to the processing cost of node v. The second part reflects the cost of input data transfer from the source node(s) of project r to the computing node v. The third element is the cost of output data transfer. In similar way we define the bandwidth consumed in the network due to allocation ω_{vr}

$$g(\omega_{vr}) = (1 - s_{vr})a_r + \textstyle\sum_{w \neq v} t_{wr}b_r \qquad (9)$$

Notice that if node v is the source node of project r ($s_{vr} = 1$), then the bandwidth consumption related to the input data transfer is 0. The proposed representation of the problem is equivalent to the linear programming formulation given by (1)-(7), however it is more convenient for construction the GRASP algorithm.

The Mixed Objective Function (MOF) is introduced to tackle the problem of fulfilling the constraints (3) and (4), especially for heavily loaded networks. To formulate this function, we first normalize functions $f(\omega_{vr})$ and $g(\omega_{vr})$. Let f^{min} and f^{max} denote the minimum and maximum value of $f(\omega_{vr})$ taking into account all possible allocations ω_{vr} $v = 1,2,\ldots,V$ and $r = 1,2,\ldots,R$. Analogously we define g^{min} and g^{max} as the minimum and maximum value of $g(\omega_{vr})$ over all possible allocations ω_{vr}. Normalized functions have the value in the range [0,1] and are formulated in the following way

$$f^{\text{norm}}(\omega_{rv}) = \frac{f(\omega_r) - f^{\min}}{f^{\max} - f^{\min}} \quad g^{\text{norm}}(\omega_{rv}) = \frac{g(\omega_r) - g^{\min}}{g^{\max} - g^{\min}} \tag{10}$$

The MOF of allocation ω_{vr} is formulated as follows, where $\beta \in [0,1]$ is a tuning parameter to regulate the tradeoff between the cost and the bandwidth functions.

$$MOF(\omega_{vr}) = \beta f^{\text{norm}}(\omega_{vr}) + (1 - \beta)\, g^{\text{norm}}(\omega_{vr}) \tag{12}$$

The algorithm is divided into two parts: solution finding and parameter adaptation. The solution is found in the following way. Algorithm calculates value of $MOF(\omega_{vr})$ for every allocation ω_{vr}. The Ranking List (RL) is created by sorting all the allocations descending by the value of function (12). Then, the Restricted Candidate List (RCL) is created. It contains all allocations, for which $MOF(\omega_{vr}) + \alpha \geq MOF^{\max}$, where $MOF^{\max}$ denotes the maximum value of $MOF(\omega_{vr})$ over all possible allocations. Parameter α is another tuning parameter of the algorithm used to find the tradeoff between random search and greedy search (if $\alpha = 1$ then it's a pure random search, when $\alpha = 0$ it is a greedy search). The algorithm chooses an allocation randomly from the RCL and tries to add it to the solution. If it is impossible due to constraints, the allocation is removed from the RCL and RL. If the operation is successful and if the project has no further tasks to allocate all allocations concerning the project are removed from RL and RCL. If RCL becomes empty (for example α parameter was set with such a low value, that RCL contained only one or few elements) the new RCL is created, using the distance of value of MOF to the best value of MOF in remaining elements in RL. If RL is empty, the procedure ends. If RL is empty and not all tasks have been allocated on nodes, the module returns incomplete solution.

The algorithm is multi–start, i.e. the user can define the number of rounds to perform – RR and the number of solutions to find in each round – RN. After each round, the performance of algorithm is evaluated, parameters α and β are modified using user defined fixed parameters α_{change} and β_{change}, which denote the elementary change of parameters. If the solution is found, then either α or β is randomly selected to be decreased. If the average cost of solutions in this round is greater than from the previous round, the algorithm further decrements one of the parameters. Otherwise, the parameter is increased.

5 Results

The main goal of the numerical experiments was to evaluate the GRASP algorithm in comparison against optimal results yielded by CPLEX 11.0 solver [9]. Several network computing systems consisting of 50 computing nodes and project sets including 30 projects were created randomly. In general we made 66 tests. Among these 66 tests we selected three cases for tuning of parameters α and β: easy case – in which the network load was low and solution was relatively simple to find (average result 2.69% worse than optimal), middle case (the average optimality gap 6.08%) and a hard case – for which finding an acceptable solution was relatively difficult (18.07%). For each case combinations of different α and β values from range 0 – 0.5 were studied. Fig. 1 shows the difference between cost of the GRASP solution (RN = 100, RR = 60) and optimal solution as a function of α and β values for easy and difficult case. Presented results indicate that the best solutions were usually found in the neighborhood of unacceptable solutions, mainly for low α and β parameters. This is why it was decided to set $\alpha = \beta = 0$. For more difficult cases it can mean that a longer time will be used for adaptation.

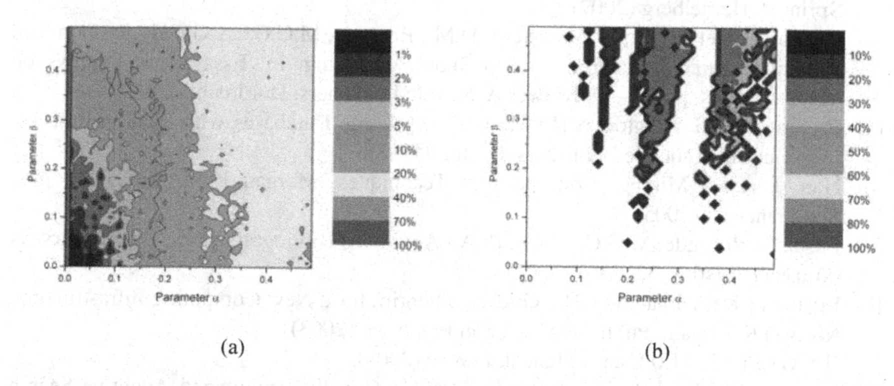

Fig. 1 The $\alpha - \beta$ space for the easy case (a) and difficult case (b)

Using the obtained tuning information we run experiments for all 66 cases. The average gap to optimal results was 2.51%. Moreover, we made tests to verify the influence of the number of rounds on the solution quality. In most cases the algorithm converged to final solution in about 50-100 rounds.

6 Concluding Remarks

In this paper we have focused on optimization of overlay Grid systems dedicated distributed MCSs. We have formulated an Integer Programming model and proposed a heuristic GRASP algorithm. We have run a large number of experiments to tune the algorithm. The obtained results show that our GRASP algorithm provides results very close to optimum. The main advantage of the heuristic approach comparing to CPLEX solver is smaller execution time. Moreover, for large problem instances the only the GRASP algorithm is able to find feasible results due to complexity of the branch-and-cut method.

Acknowledgments. This work is supported in part by The Polish Ministry of Science and Higher Education under the grant which is being realized in years 2010-2013.

References

[1] Alexandre, L.A., Campilho, A.C., Kamel, M.: Combining Independent and Unbiased Classifiers Using Weighted Average. In: Proc. of the 15th Internat. Conf. on Pattern Recognition, vol. 2, pp. 495–498 (2000)
[2] Alpaydin, E.: Introduction to Machine Learning. The MIT Press, London (2004)
[3] Biggio, B., Fumera, G., Roli, F.: Bayesian Analysis of Linear Combiners. In: Haindl, M., Kittler, J., Roli, F. (eds.) MCS 2007. LNCS, vol. 4472, pp. 292–301. Springer, Heidelberg (2007)
[4] Binato, S., Hery, W.J., Loewenstern, D.M., Resende, M.G.C.: A Greedy Randomized Adaptive Search Procedure for Job Shop Scheduling. In: Essays and Surveys on Metaheuristics, pp. 58–79. Kluwer Academic Publishers, Dordrecht (2002)
[5] Freitas, A.A., Lavington, S.H.: Mining Very Large Databases with Parallel Processing. Kluwer Academic Publishers, Boston (1998)
[6] Han, J.: Data Mining: Concepts and Techniques. Morgan Kaufmann Publ. Inc., San Francisco (2005)
[7] Festa, P., Resende, M.G.C.: GRASP: An Annotated Bibliography. Essays and surveys on metaheuristics, 325–367 (2002)
[8] Foster, I., Kesselman, C.: The Grid 2: Blueprint for a New Computing Infrastructure. Morgan Kaufmann Publishers Inc., San Francisco (2003)
[9] ILOG CPLEX 11.0 User's Manual, France (2007)
[10] Leonidas, S., Pitsoulis, L., Resende, M.G.C.: Greedy Randomized Adaptive Search Procedures. In: Handbook of Applied Optimization. Oxford University Press, Oxford (2002)
[11] Jain, A.K., Duin, P.W., Mao, J.: Statistical Pattern Recognition: A Review. IEEE Trans. on PAMI 22(1), 4–37 (2000)
[12] Kuncheva, L.I., Whitaker, C.J., Shipp, C.A., Duin, R.P.W.: Limits on the Majority Vote Accuracy in Classier Fusion. Pattern Analysis and Applications 6, 22–31 (2003)
[13] Kuncheva, L.I.: Combining pattern classifiers: Methods and algorithms. Wiley, Chichester (2004)
[14] Magoulès, F., Nguyen, T., Yu, L.: Grid Resource Management: Toward Virtual and Services Compliant Grid Computing. CRC Press, Boca Raton (2009)
[15] Nabrzyski, J., Schopf, J., Węglarz, J. (eds.): Grid resource management: state of the art and future trends. Kluwer Academic Publishers, Boston (2004)
[16] Wadenstein, M.: The LHC data stream. Nordic DataGrid Facility (2008)
[17] Walkowiak, K., Woźniak, M.: Decision tree induction methods for distributed environment. In: Men-Machine Interactions, Advances in Intelligent and Soft Computing, pp. 201–208 (2009)
[18] Zhu, Y., Li, B.: Overlay Networks with Linear Capacity Constraints. IEEE Transactions on Parallel and Distributed Systems 19(2), 159–173 (2008)

A Scatter Search Based Approach to Solve the Reporting Cells Problem

Sónia M. Almeida-Luz, Miguel A. Vega-Rodríguez, Juan A. Gómez-Pulido, and Juan M. Sánchez-Pérez

Abstract. This paper presents a new approach based on the Scatter Search (SS) algorithm, to solve the mobile Location Management problem using the Reporting Cells (RC) strategy. The RC problem is applied to achieve the best configuration of the mobile network, defining what cells should work as RC, with the objective of minimizing the costs involved. In this work we perform five distinct experiments with the aim of determining the best values for the Scatter Search parameters, when applied to the RC problem. We use 12 test networks with the objective of comparing the results achieved with those obtained through other algorithms from our previous work and by other authors. The experimental results prove that this SS based approach outperforms the results obtained by other approaches presented in the literature, which is very encouraging.

Keywords: Scatter Search, Reporting Cells Problem, Optimization, Location Management, Mobile Networks.

1 Introduction

The Location Management (LM) is an important process of the mobility management in the mobile networks. It is principally characterized by managing the network configuration, considering the users' movements and tracing, with the objective of minimizing the costs involved [1].

The LM is partitioned in two main operations, over the mobile networks: location update (LU), used to notify the current location, performed by mobile terminals when they change their location, and location paging (P) that corresponds to

---

Sónia M. Almeida-Luz
Polytechnic Institute of Leiria, School of Technology and Management,
Department of Informatics Engineering, 2400 Leiria, Portugal
e-mail: sluz@estg.ipleiria.pt

Miguel A. Vega-Rodríguez · Juan A. Gómez-Pulido · Juan M. Sánchez-Pérez
University of Extremadura, Dept. Technologies of Computers and Communications,
Escuela Politécnica. Campus Universitario s/n. 10003 Cáceres. Spain
e-mail: {mavega,jangomez,sanperez}@unex.es

E. Corchado et al. (Eds.): SOCO 2010, AISC 73, pp. 145–152.
springerlink.com

the operation of determining the location of the mobile user terminal, which is executed by the network when it tries to direct an incoming call to the user. LM has been redefined in several different strategies, which are grouped in two main groups: static and dynamic schemes [1, 2]. The Reporting Cells (RC) is one of the most common schemes, included in the static schemes (the most usual ones) that consider, for all users, the same network behavior.

This paper presents a new approach, based on the Scatter Search (SS) algorithm, to solve the reporting cells planning problem. The main objective of this problem is to optimize the configuration of mobile networks, in a way to minimize the involved costs. The paper is organized as follows. In section 2 we explain the main concepts of the RC problem and present the location management costs involved. Section 3 provides a brief description of SS algorithm. In section 4, we expose the implementation details, explain and analyze the experimental results, and then compare the results obtained with those accomplished by other authors. Finally, section 5 contains conclusions and future work.

2 Reporting Cells Planning Problem

The Reporting Cells scheme was proposed by Bar-Noy and Kessler [3] with the objective of minimizing the cost of tracking mobile users. This strategy is characterized by defining a subset of cells as reporting cells and the others as non-reporting cells (nRC), as it is possible to see in Fig. 1(a). A new location update is only performed when the mobile terminals change their location and move to one reporting cell. If an incoming call must to be routed to the mobile user, the search can be restricted to his last reporting cell known and their respective neighbors which are non-reporting cells.

For each cell in the network, it is necessary to calculate the vicinity factor, which represents the maximum number of cells that the user must page when an incoming call occurs. For detailed information on the vicinity values (shown in Fig. 1(b)) calculi, refer to [4].

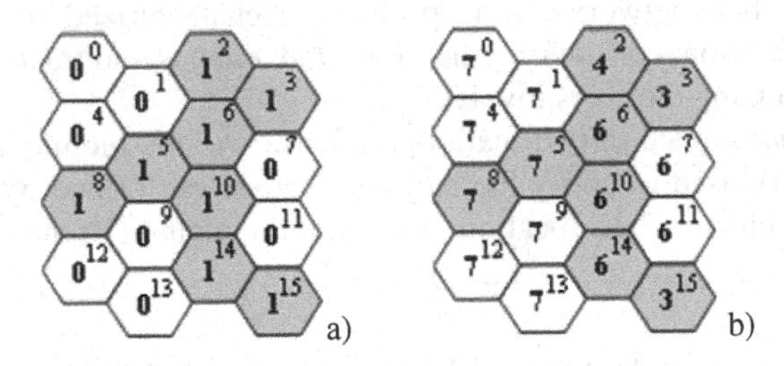

Fig. 1 Reporting Cells planning: a) nRC (0) and RC (1); b) Vicinity values

Considering earlier studies and experiments [4, 5], we have the following generic formula (2.1) used to determine the LM cost:

$$Cost = \beta \times N_{LU} + N_P . \tag{2.1}$$

The cost of location updates is given by N_{LU}, the cost of paging transactions is given by N_P, and finally β is a ratio constant used in a location update (because the LU cost is considered much higher than the P cost) relatively to a paging transaction in the network. Because of this, the cost of a location update is normally considered to be 10 times greater than the cost of paging, that is, β=10 [5].

In the RC scheme the location updates only are performed when a mobile user enters in a reporting cell and the vicinity factor of each cell must be considered. Due to this, the generic formula given by (2.1) must be readjusted and reformulated as [4, 6] (2.2):

$$Cost = \beta \times \sum_{i \in S} N_{LU}(i) + \sum_{i=0}^{N} N_P(i) \times V(i)\,. \tag{2.2}$$

Where $N_{LU}(i)$ is the total number of location updates for RC i, S indicates the subset of cells defined as RCs, $N_P(i)$ is the number of incoming calls attributed for cell i, N is the total number of cells that compound the mobile network configuration and $V(i)$ is the vicinity factor calculated for cell i. In this work, we will apply this formula with the objective of minimizing the LM costs.

3 Scatter Search Algorithm

Scatter search (SS) is an evolutionary algorithm introduced by Glover in 1977 [7]. SS is characterized by five main components [8, 9]: *Diversification Generation method*, *Improvement method*, *Reference Set Update method*, *Subset Generation method* and *Solution Combination method.* The pseudo-code of the SS algorithm is presented in Fig. 2 (see [8, 9] for more details).

```
SS Algorithm
1:  Start with Population P=Ø. Use Diversification Generation method to create a
    solution x and improve it with the Improvement method. If x ∉ P add x to P.
    Repeat this step until get PSize different solutions.
2:  Use the Reference Set Update method to create RefSet = {x^1,...,x^b} with b/2 best
    solutions and b/2 most diverse solutions (from the b/2 best solutions), of P.
3:  Evaluate the RefSet solutions and order the solutions, using their fitness function.
4:  Make NewSolution=TRUE.
5:  while (NewSolution) do
6:     Make NewSolution=FALSE
7:     Use the Subset Generation method and create all different subsets
8:     while(Exist subsets not examined) do
9:          Select a subset and label it as examined
10:         Apply the Solution Combination method to the solutions of the subset
11:         Apply the Improvement method to each new solution x obtained
12:         if (f(x) < f(x^b) and (x ∉ RefSet) then
13:                Set x^b = x and order solutions of RefSet
14:                Make NewSolution = TRUE
15:         end if
16:    end while
17: end while
```

Fig. 2 Pseudo-Code for Scatter Search algorithm

4 Experiments and Results Comparison

In this section, we describe the most significant implementation details and then we illustrate the test networks used. After that, we explain the experiments performed, the results obtained and respective analysis and conclusions. Finally, we will compare the results achieved with another approach based on Differential Evolution (DE) algorithm, and also with the results accomplished by other authors.

4.1 Implementation Details and Test Networks

The fitness function, which will be used to evaluate the total costs of location management for each solution, is defined according to equation (2.2) shown in section 2. To accomplish this study, we have selected a set of twelve test networks, based on realistic data and patterns [6], available in [10] as benchmark, and also used by us in a previous work [4]. We decided to use these twelve networks, because they represent four distinct groups, divided by size, and with the objective of comparing the results achieved with other authors.

Taking into account the implementation details of SS algorithm (Fig. 2), we use four main parameters: initial population size *PSize*; reference set size *RSSize* (divided into the size of the quality solutions *nQrs* and the size of the diversity solutions *nDrs*); probability of combination (crossover) *Cr*; and the number of iterations of local search *nLS*. In the *diversification generation* method the initial population is created considering the RC and nRC, where one of them is set to each cell with a probability of 50%. For the *improvement* method we decided to apply a local search that switch a RC with a neighbor nRC. We defined subsets of size 2, to be used in the *subset generation* method. Finally, for the combination method we implemented a crossover that might be applied to a maximum of four crossover points considering a predetermined probability.

Based on the suggestions of several authors [8, 9] we decided to start the experiments setting the following values: *PSize*=100; *RSSize*=10; *nQrs*=5; *nDrs*=5; *Cr*=0.2; *nLS*=1.

4.2 Simulation Results and Analysis

The main objective of this study is to understand the behavior of SS applied to the reporting cells planning problem.

In this section, we explain the five main experiments, which we have performed, in order to adjust the parameters of the SS algorithm. With the objective of assuring the statistical relevance of the results obtained, we have performed 30 independent runs, for each experiment, including each parameters' configuration. However, due to space reasons, we only expose the most important decisions and conclusions of all these experiments.

4.2.1 Determining the Population Size

The SS algorithm starts with an initial population of a predetermined number of distinct solutions (*PSize* parameter). Considering this, the first experiment had the objective of determining the ideal number of solutions that should compound the population. Using the twelve test networks and considering the initial values, set to each one of the other parameters, we tried *PSize* with the values 10, 25, 50, 75, 100, 125, 150, 175 and 200. Evaluating the results, we decided to proceed to the next experiment setting *PSize*=175, because, with this configuration, the best values for the best and average fitness were reached.

4.2.2 Determining the RefSet Size

After defining *PSize*, it was necessary to determine the size of the *RefSet* (*RSSize* parameter) that will include the best and most diverse solutions. Setting *PSize*=175, from the first experiment, considering the other initial parameters' values and, for the twelve test networks, we tested the following values for *RSSize*: 2, 4, 6, 8, 10, 12, 14, 16, 18 and 20. Observing the statistical results we noticed that they were improving with the increment of the *RSSize*, so we elected *RSSize*=20 to continue for the next experiment. Furthermore, we verified that the improvements with higher *RSSize* (than 20) were not significant.

4.2.3 Dividing the RefSet between nQrs and nDrs

The third experiment was defined to divide the *RefSet* between quality *nQrs* and most diverse *nDrs* solutions. Considering the *PSize*=175 and *RSSize*=20 (from the 2 earlier experiments) and fixed the other original parameters, we checked all the feasible combinations of *nQrs* and *nDrs* (considering that their sum has to be 20). Analyzing the results obtained, we concluded that it was with the division of the *RSSize* in *nQrs*=14 and *nDrs*=6 that the best values for the best and average fitness were obtained.

4.2.4 Determining the Combination Probability

In the fourth experiment we had the purpose of determining the best combination (crossover) probability *Cr* to be used in the solution combination method. So, in order to perform this experiment, we assigned *PSize*=175, *RSSize*=20, *nQrs*=14 and *nDrs*=6 from the earlier experiments, maintained the initial *nLS*=1 and executed the following configurations for *Cr*: 0.1, 0.2, 0.3, 0.4, 0.5, 0.6, 0.7, 0.8 and 0.9. After finishing these executions and evaluating the best average results, we decided to proceed with *Cr*=0.6 for the last experiment.

4.2.5 Determining the Ideal Local Search Number

The fifth and last experiment had the objective of electing the most adequate number of local search iterations (*nLS*) that should be performed in the improvement method. We have fixed the values of each parameter, reached in the previous experiments, as: *PSize*=175, *RSSize*=20, *nQrs*=14, *nDrs*=6, *Cr*=0.6 and testing the following values for *nLS*: 1, 2, 3, 4 and 5. Analyzing the results obtained, we could observe that it was with *nLS*=4 that the best (lowest) fitness cost and majority of the best average results were achieved. We also noticed that with an *nLS*=5 the evolution of the average became worst or even negative, so we select *nLS*=4 as the most adequate.

After we finished these experiments we had obtained the best configuration for the SS parameters when applied to the reporting cells problem, i.e.: *PSize*=175, *RSSize*=20, *nQrs*=14, *nDrs*=6, *Cr*=0.6 and *nLS*=4.

In addition, we have performed a statistical analysis over all the experiments executed, using the ANOVA test. We have considered a confidence level of 95% (i.e., significance level of 5% or p-value under 0.05), which means that the differences are unlikely to have occurred by chance with a probability of 95%. In Table 1 we show the results obtained, for all the experiments, by using this test, where we can understand that the fitness differences when we use distinct values for each SS parameter have been found as significant in almost all the cases.

Table 1 ANOVA analysis over SS parameters in the RCs problem, using the 12 test networks

	TN1	TN2	TN3	TN4	TN5	TN6	TN7	TN8	TN9	TN10	TN11	TN12
PSize	4.94E-1	5.01E-1	7.56E-1	1.12E-2	3.30E-1	4.65E-1	1.20E-2	3.84E-1	2.04E-2	7.98E-1	4.58E-2	8.63E-1
RS Size	<1E-15	<1E-15	<1E-15	<1E-15	<1E-15	<1E-15	<1E-15	<1E-15	<1E-15	<1E-15	<1E-15	<1E-15
Q/D	1.30E-1	2.08E-2	<1E-15	4.29E-1	4.69E-1	1.86E-4	7.71E-1	1.69E-2	3.06E-2	1.46E-1	4.05E-1	6.43E-1
Cr	4.36E-1	3.34E-2	<1E-15	3.840-7	4.9E-10	1.21E-9	9.68E-5	6.49E-5	1.09E-4	2.33E-2	4.20E-1	8.41E-3
nLS	<1E-15	<1E-15	<1E-15	9.74E-1	2.35E-1	8.04E-1	1.26E-2	2.62E-4	4.87E-1	3.91E-7	2.99E-8	3.9E-13

4.3 Comparison of Results

Once obtained the final results, we decided to compare them with a previous work where we used a DE based approach [4]. Then, we also compare the results with those achieved by Alba et al. in [6] with a Hopfield Neural Network with Ball Dropping (HNN-BD) and a Geometric Particle Swarm Optimization (GPSO). As it is possible to see in Table 2, where we show the lowest fitness costs achieved, SS always equals or surpasses (TN7, TN10, TN11 and TN12) the results obtained by the DE based approach. We also noticed that our approach have generated better solutions and surpassed the results obtained by HNN-BD and GPSO for the networks 7, 10 and 11.

Table 2 Comparison of best LM costs accomplished by different evolutionary algorithms

	TN1	TN2	TN3	TN4	TN5	TN6	TN7	TN8	TN9	TN10	TN11	TN12
SS	98,535	97,156	95,038	173,701	182,331	174,519	307,695	287,149	264,204	385,927	357,714	370,868
DE	98,535	97,156	95,038	173,701	182,331	174,519	308,401	287,149	264,204	386,681	358,167	371,829
HNN-BD	98,535	97,156	95,038	173,701	182,331	174,519	308,929	287,149	264,204	386,351	358,167	370,868
GPSO	98,535	97,156	95,038	173,701	182,331	174,519	308,401	287,149	264,204	385,972	359,191	370,868

Furthermore, with the objective of comparing our approach with other artificial life techniques, we decided to apply the best SS configuration achieved in other 3 test networks presented in [11] (also used and referred in [4, 6]). Using these 3 test networks (4x4, 6x6 and 8x8 instances) we want to compare our results with those accomplished by Genetic Algorithms (GA), Ant Colony algorithm (AC) and Tabu Search (TS). After finished these additional experiments, we realized that our approach performs well, because for all the 3 networks (with 30 runs for each one) we always obtain the same fitness, which corresponds to the lowest costs that were attained. For the Test-Network-1, we obtained the same results accomplished by the other algorithms, i.e. fitness equal to 92,833. Relatively to the Test-Network-2, we achieved a fitness of 211,278, the same as DE and TS and surpassed the GA (fitness of 229,556) and AC (fitness of 211,291). Finally for the Test-Network-3 we also got the lowest fitness cost (436,269), like with DE and surpassed the results obtained by AC (436,886) and also by GA and TS (436,283).

5 Conclusions and Future Work

In this paper we expose a new approach, based on the SS algorithm, to solve the reporting cells (RC) planning problem. We have performed five main experiments with the objective of determining the best configuration of SS parameters. Using twelve test networks, and after a big number of experiments, the most adequate values are: *PSize*=175, *RSSize*=20, *nQrs*=14, *nDrs*=6, *Cr*=0.6 and *nLS*=4.

The results achieved show that our approach can be successfully applied to the RC problem, because, comparing with the results accomplished by other authors that use GSPO and HNN+BD we obtain equal or even better fitness costs. Furthermore, when we compare the SS algorithm with other artificial life techniques as genetic algorithms (GA), ant colony algorithm (AC) and tabu search (TS), our approach shows the best performance, obtaining better solutions (the optimal ones to the best of our knowledge) for all the networks used.

As future work we plan to compare the results of the Reporting Cells strategy with the ones accomplished with the Location Areas strategy, which is another well-known static scheme of location management. With this study we want to understand which is the most adequate, since both consider the location update and paging costs of the location management.

Acknowledgments. This work was partially funded by the Spanish Ministry of Science and Innovation and FEDER under the contract TIN2008-06491-C04-04 (the M* project). Thanks also to the Polytechnic Institute of Leiria, for the economic support offered to Sónia M. Almeida-Luz to make this research.

References

1. Tabbane, S.: Location Management methods for third generation mobile systems. IEEE Communications Magazine 35, 72–84 (1997)
2. Wong, V.W.S., Leung, V.C.M.: Location management for next-generation personal communications networks. IEEE Network 14(5), 18–24 (2000)
3. Bar-Noy, A., Kessler, I.: Tracking mobile users in wireless communications networks. IEEE Transactions on Information Theory 39, 1877–1886 (1993)
4. Almeida-Luz, S.M., Vega-Rodríguez, M.A., Gómez-Pulido, J.A., Sánchez-Pérez, J.M.: Applying differential evolution to the reporting cells problem. In: International Multiconference on Computer Science and Information Technology (IMCSIT 2008), Poland, pp. 65–71. IEEE Computer Society Press, USA (2008)
5. Taheri, J., Zomaya, A.Y.: A modified hopfield network for mobility management. Wireless Communications and Mobile Computing 8, 355–367 (2008)
6. Alba, E., Garca-Nieto, J., Taheri, J., Zomaya, A.Y.: New research in nature inspired algorithms for mobility management. In: Giacobini, M., Brabazon, A., Cagnoni, S., Di Caro, G.A., Drechsler, R., Ekárt, A., Esparcia-Alcázar, A.I., Farooq, M., Fink, A., McCormack, J., O'Neill, M., Romero, J., Rothlauf, F., Squillero, G., Uyar, A.Ş., Yang, S. (eds.) EvoWorkshops 2008. LNCS, vol. 4974, pp. 1–10. Springer, Heidelberg (2008)
7. Glover, F.: Heuristics for integer programming using surrogate constraints. Decision Sciences 8, 156–166 (1977)
8. Martí, R., Laguna, M., Glover, F.: Principles of scatter search. European Journal of Operational Research 169, 359–372 (2006)
9. Laguna, M., Hossell, K.P., Martí, R.: Scatter Search: methodology and implementation in C. Kluwer Academic Publishers, Norwell (2002)
10. Test Networks: `http://oplink.lcc.uma.es/problems/mmp.html` (accessed on February 2010)
11. Subrata, R., Zomaya, A.Y.: A Comparison of three artificial life techniques for reporting cell planning in mobile computing. IEEE Transactions on Parallel and Distributed Systems 14(2), 142–153 (2003)

Fuzzy Optimization of Start-Up Operations for Combined Cycle Power Plants

Ilaria Bertini, Alessandro Pannicelli, and Stefano Pizzuti

Abstract. In this paper we present a study on the application of fuzzy sets for the start-up optimisation of a combined cycle power plant. We fuzzyfy the output process variables and then we properly combine the resulting fuzzy sets in order to get a single value in the lattice [0,1] providing the effectiveness (zero bad, one excellent) of the given start-up regulations. We tested the methodology on a large artificial data set and we found an optimum which remarkably improves the solution given by the process experts.

1 Introduction

Combined cycle power plants (CCPP) are a combination of a gas turbine and a steam turbine generator for the production of electric power in a way that a gas turbine generator generates electricity and the waste heat is used to make steam to generate additional electricity via a steam turbine. For such plants, one of the most critical operations is the start-up stage because it requires the concurrent fulfilment of conflicting objectives (for example, minimise pollutant emissions and maximise the produced energy). The problem of finding the best trade-off among conflicting objectives can be arranged like an optimisation problem. This class of problems can be solved in two ways : with a single-objective function managing the other objectives, like thermal stress, as constraints, and with a multi-objective approach.

At present, the problem of CCPP start-up optimisation has been tackled in the first way using simulators. As example, in [1] through a parametric study, the start-up time is reduced while keeping the life-time consumption of critically stressed components under control. In [11] an optimum start up algorithm for CCPP, using a model predictive control algorithm, is proposed in order to cut down the start-up time keeping the thermal stress under the imposed limits. In [4] a study aimed at reducing the start-up time while keeping the life-time consumption of the more critically stressed components under control is presented.

Ilaria Bertini · Stefano Pizzuti · Alessandro Pannicelli
Energy, New technologies and sustainable Economic development Agency (ENEA)
'Casaccia' R.C., Via Anguillarese 301, 00123 Rome, Italy
e-mail: ilaria.bertini@enea.it, stefano.pizzuti@enea.it,
alessandro.pannicelli@enea.it

E. Corchado et al. (Eds.): SOCO 2010, AISC 73, pp. 153–160.
springerlink.com

In the last decade the application research of fuzzy set theory [12] has become one of the most important topics in industrial applications. In particular, in the field of industrial turbines for energy production, it has been mainly applied to fault diagnosis [2,5,8,9,10] sensor fusion [6] and control. Particularly, in the last area in [3] it is proposed a fuzzy control system in order to minimize the steam turbine plant start-up time without violating maximum thermal stress limits. In [7] it is presented a start-up optimization control system which can minimize the start-up time of the plant through cooperative fuzzy reasoning and a neural network making good use of the operational margins on thermal stress and pollutant emissions.

In all the reported examples it is clear that the global start-up operations are not optimised. Therefore, in this work we propose an approach based on fuzzy sets in order to overcame the exposed drawbacks. Thus, for each single objective we define a fuzzy set and then we properly combine them in order to get a new objective function taking into account all the operational goals. We applied this method to a large artificial data set of different start-up conditions and we compared the best solution we found with the one given by the process experts.

The paper is structured as follows : in section 2 we describe the problem we are dealing with, section 3 reports the details of the method we carried out, section 4 shows experimental results and finally section 5 draws the conclusions.

2 The Combined Cycle Power Plants

Gas and steam turbines are an established technology available in sizes ranging from several hundred kilowatts to over several hundred megawatts. Industrial turbines produce high quality heat that can be used for industrial or district heating steam requirements. Alternatively, this high temperature heat can be recovered to improve the efficiency of power generation or used to generate steam and drive a steam turbine in a combined-cycle plant. Therefore, industrial turbines can be used in a variety of configurations:

- Simple cycle (SC) operation which is a single gas turbine producing power only
- Combined heat and power (CHP) operation which is a simple cycle gas turbine with a heat recovery heat exchanger which recovers the heat in the turbine exhaust and converts it to useful thermal energy usually in the form of steam or hot water
- Combined cycle (CC) operation in which high pressure steam is generated from recovered exhaust heat and used to create additional power using a steam turbine (fig.2.1)

The last combination produces electricity more efficiently than either gas or steam turbine alone because it performs a very good ratio of transformed electrical power per CO_2 emission. CC power plants are characterized as the 21st century power generation by their high efficiency and possibility to operate on different load conditions by reason of the variation in consumer load. CC plants are highly complex systems but being available highly powerful processors and advanced

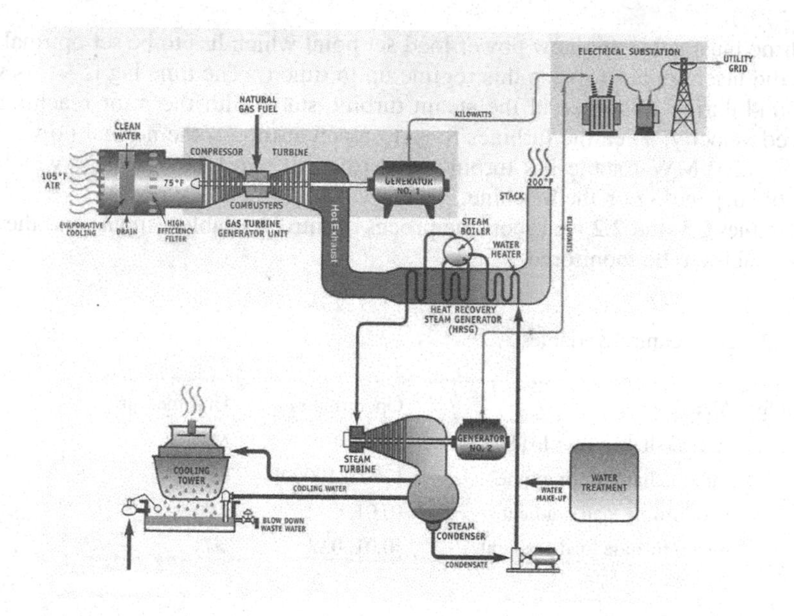

Fig. 2.1 Combined cycle power plant

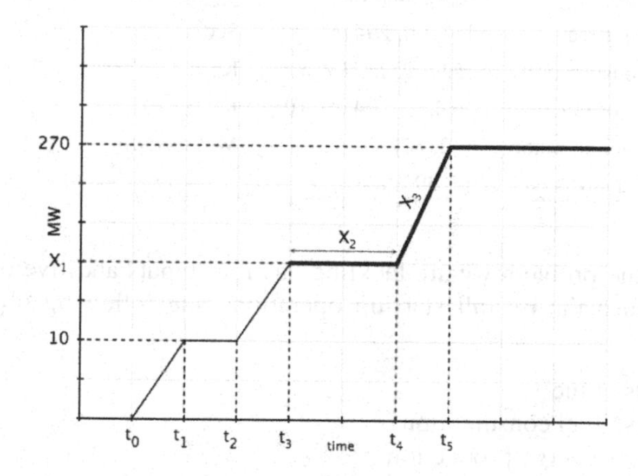

Fig. 2.2 Combined cycle power plant start-up operations

numerical solutions, there is a great opportunity to develop high performance simulators for modelling energy systems in order to consider various aspects of the system.

The start-up scheduling is as follows (fig.2.2). From zero to time t_0 (about 1200 sec) the rotor engine velocity of the gas turbine is set to 3000rpm. From time t_0 to t_1 the power load is set to 10 MW and then the machine keeps this regime up to time t_2. All this initial sequence is fixed. From time t_2 to t_3 (about 3600 sec) the machine must achieve a new power load set point which has to be set optimal and then the machine has to keep this regime up to time t_4. The time lag $t_4 - t_3$ is variable and during this interval the steam turbine starts with the rotor reaching the desired velocity. Then the turbines have to reach at time t_5 the normal power load regime (270 MW for the gas turbine) according to two load gradients which are variable depending on the machine.

In tables 2.1 and 2.2 we report the process control variables (input) and the output variables to be monitored.

Table 2.1 Input control variables

Variable	Meaning	Operating range	Unit measure
X1	Intermediate power load set point	[20, 120]	MW
X2	Intermediate waiting time	[7500, 10000]	Sec
X3	Gas turbine load gradient	[0.01, 0.2]	MW/s
X4	Steam turbine load gradient	[0.01, 0.2]	%/s

Table 2.2 Output variables

Variable	Meaning	Operating range	Unit measure
Y1	start-up time	[11700, 29416]	Sec
Y2	fuel consumption	[53000, 230330]	Kg
Y3	energy production	[$6.45*10^8$, $4.56*10^9$]	KJ
Y4	Pollutant emissions	[12, 32]	$Mg*sec/Nm^3$
Y5	thermal stress	[8, 3939]	-

Therefore, the problem we are tackling has four inputs and five outputs and in order to optimise the overall start-up operations, the following objectives need fulfilling:

- minimise time
- minimise fuel consumption
- maximise energy production
- minimise pollutant emissions
- minimise thermal stress

3 Fuzzy Sets Definition

In this paragraph we describe how, with the support of the process experts, we defined the single fuzzy sets (tab. 3.1) over the output variables (tab. 2.2) and how we composed them in order to get a cost function ranging in the lattice [0,1]. Therefore, we got an index representing the global start-up performance (0 = bad, 1 = excellent).

Table 3.1 Fuzzy sets

Fuzzy set	Membership Function $\mu_{Fi}(y_i)$	Variable	Weight w_i	t	c	Goal
F1	1-Sigmoid	Y1	0.2	8000	110000	Min
F2	1- Sigmoid	Y2	0.1	800	16200	Min
F3	Sigmoid	Y3	0.1	$0.4*10^9$	$1.8*10^9$	Max
F4	1- Sigmoid	Y4	0.3	2	25	Min
F5	1- Sigmoid	Y5	0.3	20	150	Min

Where *c* and *t* are the parameters of the sigmoid function

$$\text{Sigmoid} = \frac{1}{1+e^{\frac{c-x}{t}}} \tag{3.1}$$

Therefore, the fuzzy sets have been composed through a weighted sum and the result is a fuzzy set whose membership function is

$$\mu(y_1,y_2,y_3,y_4,y_5) = w_1\mu_{F1}(y_1) + w_2\mu_{F2}(y_2) + w_3\mu_{F3}(y_3) + w_4\mu_{F4}(y_4) + w_5\mu_{F5}(y_5) \tag{3.2}$$

This composition has been finally chosen because we found out that for this problem the intersection was too restrictive (only one objective with a low value is sufficient to severely affect the whole performance) and the union was too lazy (only one objective with a high value is sufficient to have a high global performance). Thus, we have eventually applied the weighted sum operator, which is an intermediate composition between intersection and union, which gives a global performance proportional to the optimality degree of each single objective.

The values of the weights w_i has been given by the process experts according to the importance of the corresponding objectives.

4 Experimentation

Experimentation has been carried out on artificial data obtained by means of a software simulator carried out by AnsaldoEnergia[1], a Finmeccanica company, which is the Italian leading thermoelectric power plants producer.

The operational ranges of the 4 four control variables reported in tab.1 have been discretised in 11 steps, therefore we got a data set made of $11^4 = 14641$ different start-up simulations. For each point we applied the proposed fuzzy approach and then we retrieved the point with the highest performance value.

In the following tables we compare the solution used by the experts (Exp) to the optimal one (Opt) given by the proposed approach.

[1] Reference: `ivo.torre@aen.ansaldo.it`

Table 4.1 Regulations comparison

	X1	X2	X3	X4
Exp	80	8500	0.037	0.017
Opt	90	7500	0.077	0.052

Table 4.2 Output comparison

	Y1	Y2	Y3	Y4	Y5	Performance
Exp	21070	143557	$2.5*10^9$	25	10	0.53
Opt	14400	98700	$1.6*10^9$	22	38	0.84
Nominal Variation	-38%	-25%	-22%	-15%	+0.7%	+0.31

The nominal variation of the last row is calculated as

$$\frac{Yopt_i - Y\exp_i}{\max_i - \min_i} * 100 \qquad (4.1)$$

where $Yopt_i$ is the outcome of the optimal solution, $Yexp_i$ the outcome of solution given by human experts, max_i and min_i are the maxima and minima operating range values of the five output variables (tab.2.2).

As first glance, it is clear that from these results the overall start-up performance has been remarkably improved (from 0.53 to 0.84).

In particular, the solution found cuts considerably down the start-up time (-38%), consumption (-25%) and emissions (-15%) keeping the thermal stress very low.

This solution has been actually acknowledged by the experts as the optimal balance for the start-up problem also because it is not very far (tab.4.1) from the solution provided by the them. This is important because it is reasonable the optimal solution to be near (in the input space) the one given by the experts.

As final remark we want to point out that the solution found depends on the choice (table 3.1) of the weights w_i in formula (3.2), therefore different choices of the weights will turn out into different scenarios and thus different solutions. As example, in the most trivial scenario (the one where all the w_i =0.2) the result would be the following.

Table 4.3 Optimal regulations for scenario w_i=0.2

X1	X2	X3	X4
90	7500	0.097	0.052

Table 4.4 Result for scenario w_i=0.2

Y1	Y2	Y3	Y4	Y5	Performance
13900	94600	$1.55*10^9$	22	38	0.78

5 Conclusion

In this paper we presented a study on the application of fuzzy sets for the overall optimisation of the combined cycle power plants start-up. Our method is based on the fuzzyfication of the output process variables which are afterwards properly combined in order to get a single value in the lattice [0,1] providing the effectiveness (zero bad, one excellent) of the given solution (start-up regulations).

In the problem we faced, human operators are able to optimise only one objective, the one which is the most critical (in combined cycle power plants this is the thermal stress), but the problem is multi-objective. Therefore the main novelty of the work is the proposed application of fuzzy sets in order to handle all the objectives and thus to optimise the global start-up operations.

We tested the methodology on a large simulated data set spanning all the input operating ranges and we found a solution (0.84) which remarkably improves the solution given by the process experts (0.53). The main reason for this is mainly due to the fact that the proposed approach keeps into account all the objectives and the solution found has been acknowledged by the experts to be the optimal balance among conflicting objectives.

As future work, we are willing to compare the proposed approach to multi-objective genetic algorithms.

References

1. Alobaida, F., Postlera, R.J., Epplea, B., Kimb, H.-G.: Modeling and investigation start-up procedures of a combined cycle power plant. Applied Energy 85(12), 1173–1189 (2008)
2. Bertini, I., Pannicelli, A., Pizzuti, S., Levorato, P., Garbin, R.: Rotor Imbalance Detection in Gas Turbines using Fuzzy Sets. In: SOCO 2009 - 4th International Workshop on Soft Computing Models in Industrial Applications, Salamanca, Spain (June 2009)
3. Boulos, A.M., Burnham, K.J.: A fuzzy logic approach to accommodate thermal stress and improve the start-up phase in combined cycle power plants. In: Proc. Instn. Mech. Engrs. Part B: J Engineering Manufacture, vol. 216, pp. 945–956 (2002)
4. Casella, F., Pretolani, F.: Fast Start-up of a Combined-Cycle Power Plant: a Simulation Study with Modelica. In: Proceedings 5th International Modelica Conference, Vienna, Austria, September 6-8, 2006, pp. 3–10 (2006)
5. Ganguli, R.: A fuzzy-logic intelligent system for a gas-turbine module and system-fault isolation. International Society of Air Breathing Engines, ISABE-2001-1112 (2001)
6. Goebel, K., Agogino, A.: Fuzzy Sensor Fusion for Gas Turbine Power Plants. GE Research & Development Center Technical Information Series (2001)

7. Matsumoto, H., Ohsawa, Y., Takahasi, S., Akiyama, T., Hanaoka, H., Ishiguro, O.: Startup Optimization of a Combined Cycle Power Plant Based on Cooperative Fuzzy Reasoning and a Neural Network. IEEE Transactions on Energy Conversion 12(1), 51–59 (1997)
8. Ogaji, S.O.T., Marinai, L., Sampath, S., Singh, R., Prober, S.D.: Gas-turbine fault diagnostics: a fuzzy-logic approach. Applied Energy 82(1), 81–89 (2005)
9. Priya, A., Riti, S.: Gas Turbine Engine Fault Diagnostics Using Fuzzy Concepts. In: AIAA 1st Intelligent Systems Technical Conference, Chicago, Illinois (2004)
10. Siu, C., Shen, Q., Milne, R.: TMDOCTOR: a fuzzy rule and case-based expert-system for turbomachinery diagnosis. SAFEPROCESS. In: Proceedings of IFAC symposium, vol. 1, pp. 556–563 (1997)
11. Tetsuya, F.: An optimum start up algorithm for combined cycle. Transactions of the Japan Society of Mechanical Engineers 67(660), 2129–2134 (2001)
12. Zimmerman, H.J.: Fuzzy set theory. Kluwer Academic, Boston (1991)

Catalog Segmentation by Implementing Fuzzy Clustering and Mathematical Programming Model

Amir Hassan Zadeh, Hamed Maleki, Kamran Kianfar,
Mehdi Fathi, and Mohammad Saeed Zaeri

Abstract. This work is concerned with the fuzzy clustering problem of different products in j variant catalogs, each of size i products that maximize customer satisfaction level in customer relationship management. The satisfaction degree of each customer is defined as a function of his/her needed product number that exists in catalog and also his/her priority. To determine the priority level of each customer, firstly customers are divided to three clusters with high, medium and low importance based on his/her needed products list. Then, all customers have been ranked based on their membership level in each of the above three clusters. In this paper in order to cluster customers, fuzzy c-means algorithm is applied. The proposed problem is firstly modeled as a bi-objective mathematical programming model. The objective functions of the model are to maximize the number of covered customers and overall satisfaction level results of delivering service. Then, this model is changed to a single integer linear programming model by applying fuzzy theory concepts. Finally, the efficiency of the proposed solution procedure is verified by using a numerical example.

Keywords: Catalog Segmentation, Customer Clustering, Fuzzy C-means Algorithm.

1 Introduction

Nowadays, one of the business and marketing organizations goals is customer attraction. Catalog design has known as the most common way in customer attraction,

---

Amir Hassan Zadeh · Mehdi Fathi
Department of Industrial Engineering, Amir Kabir University of Technology, Tehran, Iran

Hamed Maleki
Department of Industrial Engineering, Azad University, South Tehran Branch, Tehran, Iran

Kamran Kianfar
Department of Industrial Engineering ,Isfahan University of Technology, Isfahan, Iran

Mohammad Saeed Zaeri
Iran Helicopter Support & Renewal Company (IHSRC), Tehran, Iran

E. Corchado et al. (Eds.): SOCO 2010, AISC 73, pp. 161–169.
springerlink.com

satisfaction, and retention cycle of customer relationship management [9]. Meanwhile one of the challenges that companies using catalogs face is the optimization problem of available products; this means by the catalog which products are covered and for which cluster of customers is designed. The catalog optimization has an effective role in satisfying customers' requirements and company profitability. With the fast growing number of products suggested by retailers and electronic retailers alike, the design of one catalog that can include all products is not possible. In many cases, some customers are attracted by only a small portion of the products the company carries. Nowadays, the catalog design has changed as one of the competitive tools for marketing among companies. The latest Benchmark Survey on important Issues and Trends conducted by Catalog Age magazine in 2003 indicates that catalog companies that participated in the survey spent a mean 26% of their profits on marketing expenses and that print and postage costs account for nearly half of those expenses [1]. The number of catalogs that companies send to customers and potential buyers is also growing up at a fast pace [4].

Efficiency and profitability of one catalog is evaluated by the number of available products of interest to customers that companies send to them. The companies seek optimum decision making on the customers clusters, not optimal decision making as individual [11]. In this problem, the need of a customer is satisfied if at least the specified minimum number of products of interest to him/her is included in one of the catalogs.

The classical catalog segmentation problem was introduced by Kleinberg et al. [7]. This problem consists of designing j catalogs, each of size i products that maximize the number of covered customers. This paper in different from previous studies in the way of defining the problem and also the solution method. Xu et. al. [12] tried to develop j catalogs with i products in order to maximize the overall number of catalog purchased. In his Model, the interest constraint is minimized and the profit constraint is maximized so that the profit of products purchased by customers who have at least t interesting products in receiving catalogs is maximized. Amiri [1] presented a mathematical model that consists of designing j catalogs, each of size i products and the target is finding the maximum proportion of costumers covered by catalogs and maximizes the number of covered customers. But it is unrealistic to assign a crisp value for a subjective issue, especially when the information is vague or imprecise [5]. This is a main motivation of this study. In this paper, the authors modeled the catalog design problem as a fuzzy mathematical programming model and this is the main contribution of this research. Other contribution of this research is considering of priority degree of customers. Efficiency of the proposed model has been verified through a numerical example. This method is quite novel in the area of catalog segmentation problems. In this model, we assume that customers and their product interests are identified. The key input data in order to implement catalog segmentation is the customer interest database that consists of the set of products which each customer is attracted in. The product interests of a customer can be gained either by aggregating all purchase transactions of the customer or by obtaining his/her explicit preferences from a set of products.

The remainder of the paper is organized as follows. In Section 2, we give the definition and formulation of the problem. Then, circumstance of the priority degree determination on each customer and in section 4 clustering algorithm of *fuzzy c-means* is described briefly. In section 5, a bi-objective mathematical programming model is proposed. This model has been changed to a single integer linear programming model by applying fuzzy theory concepts in section 6. Afterwards, a numerical example has been used for better understanding of the problem. Finally, section 8 has been addressed to present a general conclusion of the paper.

2 Problem Definition

Assume that customers and their product interests are known. The main input data for catalog segmentation is the customer interest database that contains the set of products that each customer is interested in. The product interests of a customer can be obtained either by aggregating all purchase transactions of the customer or by obtaining his/her explicit preferences from a set of products. The customer-oriented catalog segmentation problem consists of designing j catalogs, each of size i products that satisfy the defined objective function of the model at the best possible circumstance. It's assumed the need of a customer is satisfied if at least the specified minimum number (t) of the products of interest to him/her is in one of the catalogs. In general, the customer catalog segmentation problem is a customer clustering problem where the task is to determine j clusters of customers where each cluster is defined by a set of interested products to the customers and each customer is apportioned to a catalog that contains the largest number of products of interest to him/her [1],[3]. An appropriate clustering should meet two restrictions which each cluster should contain of i products and the needs of a customer can be covered only by a catalog which contains at least i products of interest to the customer.

In this paper, the customers fuzzy clustering and the catalog segmentation problem are modeled as a bi-objective programming model and then are solved. In this model, first objective function shows the number of customers that we want to satisfy their needs through the catalogs set and second objective function implies the marginal utility results of allocating the catalogs to the customers. In order to build second objective function, at first it's necessary that a priority degree is defined per customer.

3 Determining Priority Degree of Each Customer

In order to determine the priority degree of the customers, initially they are divided into three clusters with high, medium and low importance based on their needed products list. Because of fuzzy nature of the problem and the fact that each of the customers may partially belong to each of three clusters noted above, *fuzzy c-means* algorithm has been used for clustering the customers. More details related to this algorithm have been addressed in section 4. If m_{ij} show the membership degree of the customer i in cluster j and c_j show the importance coefficient of cluster j, the priority degree of the customer i will be calculated as the following equation:

$$p_i = \sum_{j=1}^{3} m_{ij} . c_j \tag{1}$$

4 Fuzzy c-Means Clustering Algorithm

Fuzzy c-means (FCM) is a method of clustering which allows one piece of data to belong to two or more clusters [2]. This method is frequently used in pattern recognition. It is based on minimization of the following objective function:

$$J_m = \sum_{i=1}^{N}\sum_{j=1}^{C} u_{ij}^{m}\left\|x_i - c_j\right\|^2 \qquad 1 \le m < \infty \tag{2}$$

Where m is any real number greater than 1, u_{ij} is the membership degree of x_i in (delete the) cluster j, x_i is the ith of d-dimensional measured data, c_j is the d-dimension center of the cluster [6]. Fuzzy partitioning is carried out through an iterative optimization of the objective function shown above, with the update of membership u_{ij} and the cluster centers c_j by:

$$c_j = \frac{\sum_{i=1}^{N} u_{ij}^{m} . x_i}{\sum_{i=1}^{N} u_{ij}^{m}} \tag{3}$$

$$u_{ij} = \frac{1}{\sum_{k=1}^{C}\left(\frac{\left\|x_i - c_j\right\|}{\left\|x_i - c_k\right\|}\right)^{\frac{2}{m-1}}} \tag{4}$$

This iterative process will stop when $\| U^{(k+1)} - U^{(k)} \| < \delta$.

5 Proposed Mathematical Programming Model

The following variables and indices are used in mathematical model definition.

A. Sets and indices

- k — *Customers' index k=1, 2, …, N*
- i — *Products' Index i=1, 2, …, M*
- J — *Catalogs' index j=1, 2, …, T*

B. Model parameters

- r — *Number of available products on each catalog*
- t — *Minimum customer interest threshold (product number)*
- p_i — *Priority degree of customer i^{th}*

C. Decision variables

- x_k — *1 if catalog k is chosen, otherwise 0.*
- v_{kj} — *1 if need of customer k is covered by catalog j, otherwise 0.*
- y_{ji} — *1 if catalog j includes product i, otherwise 0.*
- a_{ki} — *1 if customer k needs to product i, otherwise 0.*

The mathematical programming model is presented as follows:

$$Max\ Z_1 = \sum_{k\in N} X_k \tag{5}$$

$$Max\ Z_2 = \sum_{k\in N} p_k X_k \tag{6}$$

$$ST.\ \sum_{i\in M} Y_{ji} = r \quad \forall j \in T \tag{7}$$

$$tV_{kj} \leq \sum_{i\in M} a_{kj} Y_{ji} \quad \forall k \in N, j \in T \tag{8}$$

$$X_k \leq \sum V_{kj} \quad \forall k \in N \tag{9}$$

$$X_k, Y_{ji}, V_{kj} \in \{0,1\} \quad \forall k \in N, j \in T, i \in M \tag{10}$$

The existing relations in the model can be described as follows:

Objective function (Z_1) maximizes the number of customers covered by the catalogs. Objective function (Z_2) maximizes overall satisfaction level results of the catalogs. Constraints (7) ensure that each catalog includes *r* products. Constraints (8) guarantee that the need of each customer should be satisfied by at least one catalog. Constraints (9) state that if the customer need is satisfied by a catalog, this catalog needs to be chosen. Constraints (10) impose the binary nature on sets of decision variables. Note that if the sizes of the catalogs are different, then the right-hand sides of constraints (7) should be set. As following, the above multi-objective mathematical programming model will be changed to a single **ILP** model by applying decision making methods based on fuzzy theory.

6 Fuzzy Multi-objective Decision Making (FMODM)

In the recent decades, Fuzzy Multi-objective Decision Making (FMODM) methods are considered by researchers. In these methods, instead of using an optimization criterion, several optimization criteria are used. Two major areas have evolved, both of which concentrate on decision making with several criteria: Multi Objective Decision Making (MODM) and Multi Attribute Decision Making [13]. In order to model a multi objective problem on fuzzy method, function h_i can be defined as follows:

$$h_i = (f_i(x)) = 1 - \frac{M_i - f_i(x)}{M_i - m_i} \qquad h_i : R \rightarrow [0,1] \tag{11}$$

In this function, value of $h_i(t)$ implies the satisfaction degree of the decision maker from function *i*th for a specific value *t*. M_i and m_i respectively represent maximum and minimum values of the function on space of the variable decisions.

To obtain values of M_i and m_i, the problem should be solved by considering only objective function *i*th and model constraints. Following figure shows the satisfaction degree of objective function *i*th.

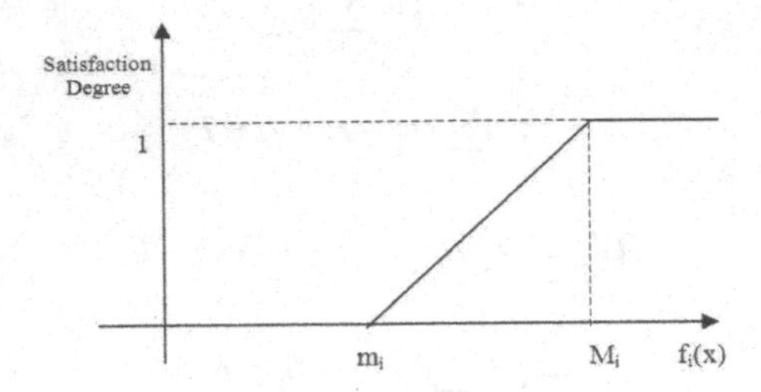

Fig. 6.1 Utility function related to decision variables

Another one of applied functions that also used in this paper is as follows:

$$h_i(t) = \begin{cases} 1 & t \geq R_i \\ u_i(t) = 1 - \dfrac{R_i - f_i(x)}{R_i - r_i} & r_i \leq t < R_i \\ 0 & t < r_i \end{cases} \tag{12}$$

Where

$R_i \leq M_i = \max\{f_i(x)\} \quad x \in X$

$r_i \geq m_i = \min\{f(x)\} \quad x \in X$

r_i, R_i respectively represent at least satisfaction level and perfect satisfaction level of objective function *i*th. This is shown in Fig. 2:

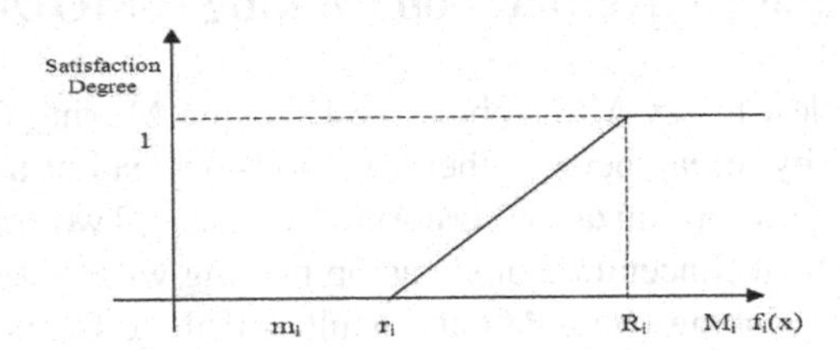

Fig. 6.2 Revised utility function

In the next stage, the objective functions are changed to a unique utility function by using one of integration operators. In this paper, standard integration operator of *max-min* is used. We should point out that instead of the operator *max-min*, each other combination of *T-norms* and *S-norms* can be used. The multi-objective mathematical programming model can be changed to an ILP model as follows:

$$
\begin{aligned}
&Max \quad \alpha \\
&s.t. \quad \alpha \le 1 - \frac{R_i - \sum_{k \in N} X_k}{R_i - r_i} \\
&\alpha \le 1 - \frac{R_i - \sum_{k \in N} p_k . X_k}{R_i - r_i} \\
&\alpha \le 1 \\
&\sum_{i \in M} Y_{ji} = r \qquad \forall j \in T \\
&tV_{kj} \le \sum_{i \in M} a_{kj} Y_{ji} \qquad \forall k \in N, j \in T \\
&X_k \le \sum V_{kj} \qquad \forall k \in N \\
&X_k, Y_{ji}, V_{kj} \in \{0,1\} \qquad \forall k \in N, j \in T, i \in M
\end{aligned} \tag{13}
$$

7 Numerical Example

A supplier with 10 different products and 15 customers is considered. The sale price for each of these products is presented in table 7.1. In table 7.2. the products of interest to each customer and in addition value summation of the products of interest to him/her are shown.

Table 7.1 The prices of available products

Product	1	2	3	4	5	6	7	8	9	10
Price	4100	6200	5400	4700	5800	6900	4000	5200	3900	5000

Table 7.2 The products of interest to each customer and their value summation

Customer	Interest products	Value of interest products	Customer	Interest products	Value of interest products
1	1,10	9100	9	1,4,6,7,9	23600
2	2,4	10900	10	2	6200
3	1,3,4,5,7	24000	11	4,9	10600
4	1,2,3,4,6,7,9,10	40200	12	1,2,5,7	20100
5	2,9	10100	13	4,5,10	15500
6	1,2,4,6,9	25800	14	5,6,7	16700
7	3,5	11200	15	1,2,3,7	19700
8	5,6,7,8	21900			

After clustering the customers by fuzzy c-means, the membership degree of each customer on each of three clusters was obtained as shown table 7.3. In the next stage, the priority of each customer was calculated according to (1) as table 7.4.

Table 7.3 The membership degree of the customers in each group

Customer	1	2 3	4 5 6	7	8 9	10	11	12	13	14	15
Low	0.99	1 0.02	0 1 0.06	0.99	0 0.02	0.93	1	0.03	0.58	0.39	0.05
Medium	0.01	0 0.96	0 0 0.88	0.01	1 0.98	0.06	0	0.96	0.39	0.59	0.94
High	0	0 0.02	1 0 0.07	0	0 0.01	0.01	0	0.01	0.03	0.03	0.01

Table 7.4 The priority degree of each customer

Customer	1	2	3	4 5 6	7	8 9	10	11	12	13	14	15
Priority degree	1.04	1.02	4.98	9 1 5.04	1.04	5 4.98	1.35	1	4.9	2.77	3.58	4.85

Assume that we want to design three catalogs which each of them include 5 different products. The results of solving the ILP model is given as table 7.5.

Table 7.5 The results of solving the integer linear programming model

Catalog number	Available products in catalog
1	1-2-4-6-9
2	2-5-6-7-8
3	1-3-4-5-7

If it's given that at least necessary utility level for satisfying the customers need equals 75% of products needed by each customer, according to catalogs achieved from solving the mathematical programming model, the need of 13 customers are satisfied. Note the marginal utility degree is equal to 82%. To solve the above model, a *PC* with processor *P4-2.4 GHz* was used and integer linear programming problem was modeled and solved by using *Lingo optimization software*.

8 Conclusion

In the paper, we have studied the fuzzy clustering problem of different products in k variant catalogs, each of size r products that maximize the customer satisfaction level in customer relationship management. The proposed problem is firstly modeled as a bi-objective mathematical programming model. Then this model is changed to a single integer linear programming model by applying the fuzzy theory concepts. Finally, the efficiency of the proposed solution procedure is examined by using a numerical example.

References

[1] Amiri, A.: Customer-oriented catalog segmentation: Effective solution approaches. Decision Support Systems 42, 1860–1871 (2006)
[2] Berget, I., et al.: New modifications and applications of fuzzy C-means Methodology. Computational Statistics & Data Analysis 52, 2403–2418 (2008)

[3] Bradley, P.S., Fayyad, U.M., Mangasarian, O.L.: Mathematical programming for data mining: Formulations and challenges. Journal on Computing 11, 217–238 (1999)
[4] Chiger, S.: Benchmark survey on critical issues and trends Catalog, pp. 32–37 (2003)
[5] Hsu, C.-C., Chen, Y.-C.: Mining of mixed data with application to catalog marketing. Expert Systems with Applications 32, 12–23 (2007)
[6] Ester, M., Ge, R., Jin, W., Hu, Z.: A Microeconomic Data Mining Problem: Customer-Oriented Catalog Segmentation. In: Proceedings of the 2004 ACM SIGKDD international conference on Knowledge discovery and data mining, Seattle, Washington, pp. 557–562 (2004)
[7] Kleinberg, J., Papadimitriou, C., Raghavan, P.: Segmentation problems. In: Proceedings of the Thirtieth Annual ACM Symposiumon Theory of Computing, pp. 473–482 (1998)
[8] Lin, C., Hong, C.: Using customer knowledge in designing electronic catalog. Expert Systems with Applications 34, 119–127 (2008)
[9] Ngai, E.W.T., et al.: Application of data mining techniques in customer relationship management: A literature review and classification. Expert Systems with Applications (2009)
[10] Rygielski, C., Wang, J.C., Yen, D.C.: Data mining techniques for customer relationship management. Technology in Society 24, 483–502 (2002)
[11] Turkey, M.: A mixed-integer programming approach to the clustering problem with an application in customer segmentation. European Journal of Operational Research (2006)
[12] Xiujuan, X., Liu, Y., et al.: Catalog segmentation with double constraints in business. Pattern Recognition Letters 30, 440–448 (2009)
[13] Zimmermann, H.J.: Fuzzy Set Theory and Its Applications, 3rd edn. Kluwer academic publishers, Dordrecht (1996)

Multi-Network-Feedback-Error-Learning with Automatic Insertion

Paulo Rogério de Almeida Ribeiro, Areolino de Almeida Neto,
and Alexandre César Muniz de Oliveira

Abstract. This work is devoted to present a control application in an industrial process of iron pellet cooking in an important mining company in Brazil. This work employs an adaptive control in order to improve the performance of the conventional controller already installed in the plant. The main strategy approached here is known as Multi-Network-Feedback-Error-Learning (MNFEL). The basic idea in MNFEL is the progressive addition of neural networks in the Feedback-Error-Learning (FEL) scheme. However, this work brings innovation by proposing a mechanism of automatic insertion of new neural networks in MNFEL. In this work, due to the unknown mathematic model of the iron pellet cooking, the plant is simulated by a previously learned neural model. In such simulation environment, the proposed method is compared against conventional PID, FEL and MNFEL.

1 Introduction

Vale is one of the world's largest mining companies with operations in production and trade of iron ore, iron pellets, nickel, coal, bauxite and others [13]. Vale is a company present in five continents which demand high quality products. Its plant in São Luís, Brazil has a pelletizing plant. This plant adds some substances to the iron ore and then puts it in pellet form. At this stage the pellets do not have consistent structure, needing a final cooking. The process of cooking is performed by 21 burner groups fed with oil, each one controlled by a Proportional-Integral-Derivative (PID) controller. In general, they perform good, except after stoppage or resumption of the process. In this case sometimes the automatic control is turned off, doing a manual action.

Paulo Rogério de Almeida Ribeiro · Areolino de Almeida Neto ·
Alexandre César Muniz de Oliveira
Universidade Federal do Maranhão (UFMA), Av. dos Portugueses, SN, Campus do Bacanga, Bacanga, 65085 – 580, São Luís, MA, Brazil
e-mail: pauloribeiro1000@yahoo.com.br, areolino@ufma.br, acmo@deinf.ufma.br

E. Corchado et al. (Eds.): SOCO 2010, AISC 73, pp. 171–178.
springerlink.com

Vale needs to improve the control, however it does not want to remove the pre-existing controller, PID is reliable and the removal increases costs, thus the solution is to improve the control system. One strategy very interesting for this purpose is called Feedback-Error-Learning (FEL) [4], it adds a subsystem in the control loop system, thereby improving the performance.

The FEL improves the performance of control system adding a Neural Network (NN) in a closed loop system [7, 12]. The FEL has an adaptive characteristic [8], i.e. a NN is added to the control system and acts in harmony with the pre-existing controller. The NN learns the inverse model of the controlled object [4]. FEL has been widely used [7, 8, 12, 14].

Whenever the network's error is not decreasing, one solution for example: restart the train or change the learning rate. However, it spends a long time and not ensures the successful outcome. But [1] proposed the use of multiples neural network, so when the network's error does not decreases, other NN is added. The strategy proposed in [1] for control is called Multi-Network-Feedback-Error-Learning (MN-FEL), it is an improvement on FEL control strategy [2, 3].

This work is devoted to supply the MNFEL strategy with automatic insertion of NN. That is very interesting because in original MNFEL the specialist knows the best moment of add a new NN, however, this can take many time and require the knowledge of the system. This pelletizing plant control has been explored in [9, 11, 10] with FEL, PI-Fuzzy and MNFEL. In this paper the results with PID, FEL, MNFEL and MNFEL with automatic insertion are compared.

Using FEL, the performance of control system is improved [7], MNFEL improves the results obtained with FEL [2] and the MNFEL's results with automatic insertion are compared with all strategies. Among the 21 burner groups, the burner group number 8 has been chosen for the experiments. It is believed that the results can be easily extended to the others groups.

This paper is organized as follows. In Section 2, the pelletizing process is shown, in order to better understand the production of pellets by the company. In Section 3, the theoretical aspects of PID, FEL and MNFEL are presented. Section 4 brings computational results, as well as a comparison between the control approaches. Finally, Section 5 presents the conclusion and directions for further works.

2 The Pelletizing Process

The pelletizing process is the agglomeration of ultrafine particles of iron ore and additives, including lime, bentonite and other inputs for the production of pellets. One of the important uses of the pellets is in the production of steel.

The production of iron pellets in Vale's pelletizing plant in São Luís, Maranhão, Brazil can be summarized by Figure 1. This is analyzed by all stages. After pellets are fired, they are ready and will be marketed around the world.

The cooking of the pellet in Figure 1 is done in an indurate machine, the pellets pass through the 21 burner groups. The temperatures are between 825.6 °C and 1350

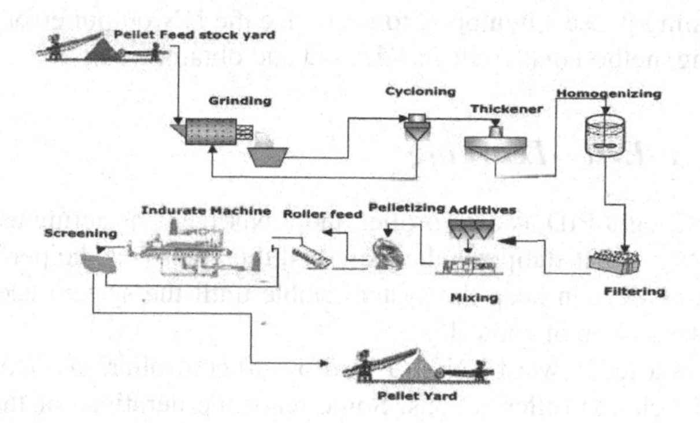

Fig. 1 Pelletizing process flowchart (Adapted from [13])

°C. The fuel used to cook the pellets is mineral oil, whose purpose is to produce heat. The cooking of the pellets is controlled by PID.

3 Control Strategies

3.1 Conventional Feedback Controller

One controller more used in industrial processes is the PID. The PID controller is defined by so called PID gains: proportional, integrative and derivative.

In the control problem presented in this paper, the output of the controller is an electrical signal, that activates a valve, expressing the level of openness (0% to 100%), through in which the oil goes. The plant's output corresponds to a value of temperature (°C).

3.2 Artificial Neural Networks

The NNs are composed by a set of artificial neurons. The neural network-based learning methods can be supervised or not [5]. The most common type of NN is the Multilayer Perceptron (MLP), where the neurons are arranged in layers.

The layers can be: Input, Hidden and Output. In this technique of machine learning, there are two phases, training and execution. The training is characterized by the modification of its parameters in order to learn the patterns of input provided during the training, while the execution phase provides outputs with no changing in its parameters for any data presented in input layer.

One of the most used algorithms for training MLP network is the so called Backpropagation [5]. Its goal is to minimize the output error function of the NN. It makes use of an input/output set, in which the network learns to do a mapping of input to output. The algorithm has two phases: forward, that produces the output of the network and backward, responsible for doing the backpropagation of error and only used in the training phase, attempting to minimize the NN output error by the gradient descending method that confronts desired and obtained output.

3.3 Feedback-Error-Learning

FEL uses a CFC, eg a PID as a controller and a NN feeds or acting as a driver in advance or feeds, so that stabilizes the plant and this improves the performance of the control. The CFC can keep the system stable until the system acquires some knowledge of the system in control.

The FEL uses a feedfoward NN as a feedfoward controller, the network online learning a feedback controller actions. Some real-time iterations of the controller process are used for the FEL neural network training, in which the strategy is to learn what actions make the PID controller stable. After training, the feedforward NN is able to drive in advance, improving the performance of the control.

This control strategy is based on studies about Central Nervous System [6]. There are a set of signals, one is called *teacher signal*, while the other signals are only input. *Teacher signal* is the error signal of NN, CFC's output. The others are input signals of network. The FEL scheme can be seen in Figure 2.

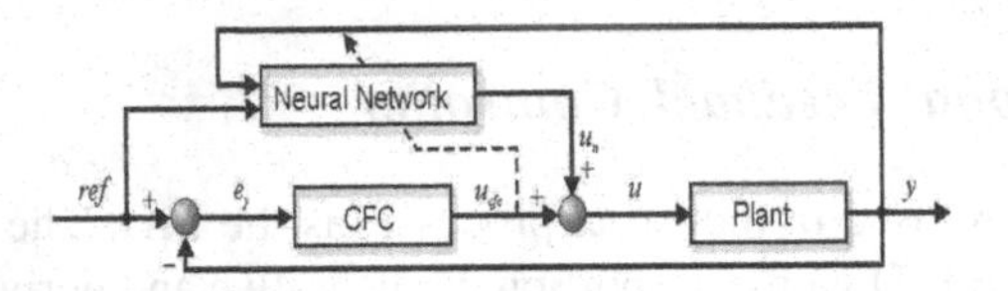

Fig. 2 Feedback-Error-Learning scheme

The weights are adjusted by backpropagation of CFC output running as output error. Minimizing it, the network indirectly minimizes the output error of the plant, and even if any component integrative, $K_i \int e(t)dt$, exists in CFC.

One can cite the advantage of the strategy that there is no removal of the existing controller, thus avoiding unnecessary costs with personnel training, the NN is easily coupled, thus, it has the adaptive characteristic [7].

3.4 Multi-Network-Feedback-Error-Learning

[1] proposed a strategy called Multi-Network-Feedback-Error-Learning (MNFEL), that is similar the FEL strategy, but instead of employing just one NN, multiple neural networks are progressively added to the control system.

The main objective of FEL strategy is acquire the inverse model of the plant, however only one NN maybe cannot, the usual approach whenever that occurs is restart the train. So, a new training is done, spending a long time, it does not ensures the successful outcome and lose the previous training. In this new training, any parameters can be modify, however, there are many parameters, as a learning rate, number of input units, hidden units, layers, input signals etc. The MNFEL strategy avoids that, because the searching to minimum can be continued with other NN.

Another advantage of MNFEL strategy it is to combine multiple NN without a coordinate, that is a hard task, the solution pointed in [1] [2] and [3] is to add only one NN a time, so only one is training in each time. These training processes drive the neural networks to follow the same way, because more than one training process, at same time, can lead to oppositions ways, i.e., each other training cancellation. The Figure 3 shows the MNFEL scheme.

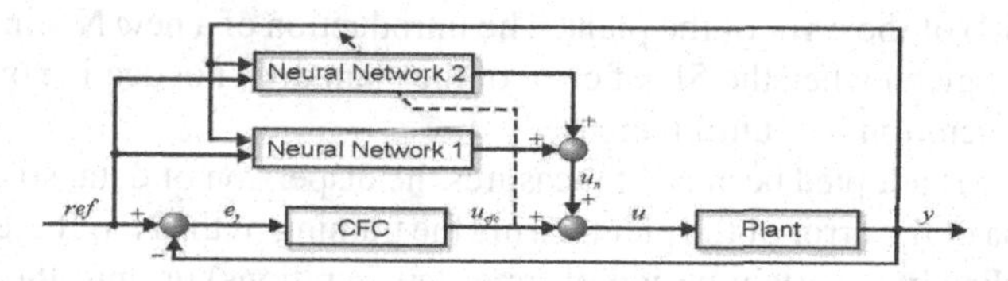

Fig. 3 Multi-Network-Feedback-Error-Learning scheme

By Figure 3 only two networks are seen, but more ones can be combined. The *teacher signal* of all is the same, CFC's output. Each network added to the control system works in cooperation with the others. That continuous training occurs because each NN has a different value of output error function, Equation 1 and 2 show the error function for first and second, respectively. They are different since they are not trained at the same time [3].

$$E(z) = z^{-M}R(z) - G(U_{cfc}(z) + U_{n1}(z)) \quad (1)$$

$$E(z) = z^{-M}R(z) - G(U_{cfc}(z) + U_{n1}(z) + U_{n2}(z)) \quad (2)$$

where $E(z)$ is the error of the plant, $R(z)$ is the set point, $G(z)$ is the transfer function of the plant, $U_{cfc}(z)$ is the control action of CFC, $U_{n1}(z)$ is the output of the first NN and $U_{n2}(z)$ is the output of the second NN.

Whenever the new NN starting in the control system the output error of the plant is different of before it be included. The aim in adding NN is to decrease the output error of the plant, starting of the point previously reached by previous NN. In the strategies that use multiple neural networks generally a coordinator component is needed. MNFEL dispenses such component, even though, in MNFEL, the networks do not conflict [2].

4 Results

The control strategies are now compared against previous Enterprise's PID controller, FEL strategy, MNFEL strategy and MNFEL with automatic insertion. The process of modeling the plant has been validated: Train and Generalization of plant's neural model [9, 10, 11]. All tests were realized in MatLab ®.

4.1 Automatic Insertion of Neural Networks

The conventional approach of using MNFEL strategy is to add manually the NN. The specialist knows the best time to introduce a new NN in the control system. However, until the arrival of this moment, a long time can be spending.

Therefore, this work presents an approach to do the automatic insertion of NN in the control system for the MNFEL strategy. The approach is to analyze the Standard Deviation (SD) of the error of the plant. The introduction of a new NN in the control system can be given when the SD of error of the plant does not decline over a period of time (few iterations) or until increase.

This metric is adopted because it measures the dispersion of data, so can analyze the dispersion of the error of the plant during the training of the NN. If the dispersion does not decline by a certain period of time (few iterations) or until the increase, it is time to add another NN.

4.2 Performace's Comparison

In that pelletizing plant control a considerable Overshoot is danger, because it burns electronics apparatus of company and a small Rise Time is required after stoppage or resumption.

The Figure 4 shows the results with automatic insertion, as a the Figure 5 shows the same NN, however, manual. The best Rise Time obtained in both insertion, automatic and manual, was MNFEL with 3 NN. The Overshoot in all strategies was not dangerous. The Figures 6 and 7 are just a zoom of the Figures 4 and 5, the purpose are show the Settling of all strategies and insertion.

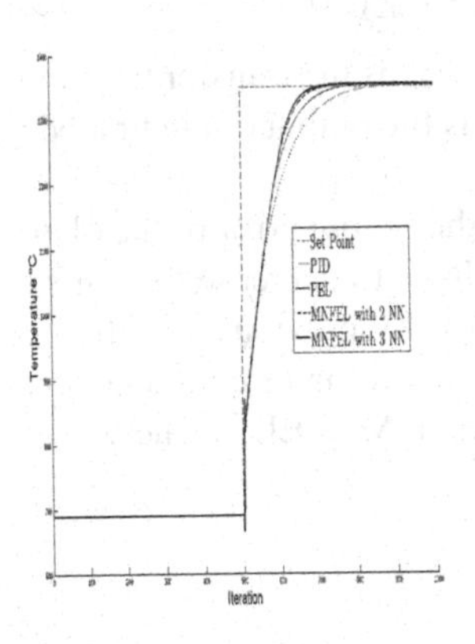

Fig. 4 Automatic

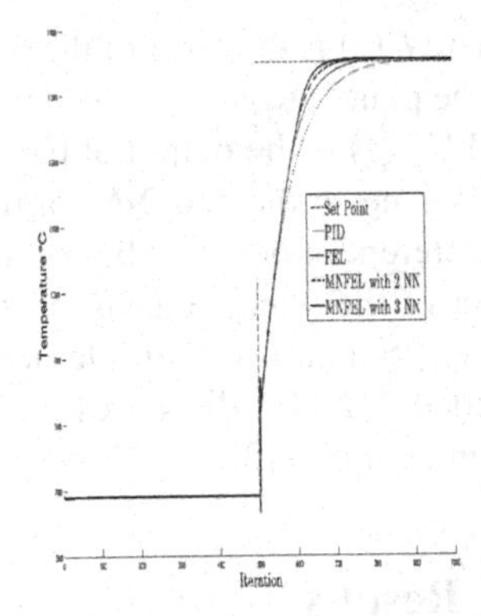

Fig. 5 Manual

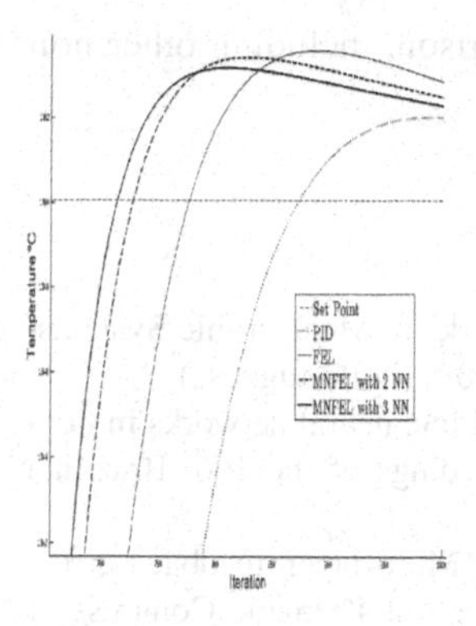

Fig. 6 Automatic - Settling

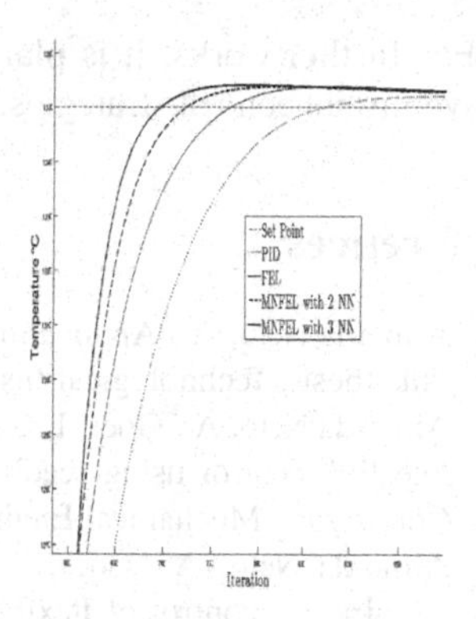

Fig. 7 Manual - Settling

The NN configurations are: $20 - 15 - 1$ (first), $40 - 15 - 1$ (second) and $60 - 15 - 1$ (third) for neurons in input, hidden and output layers, respectively. The same configurations were used in automatic and manual insertion. The network's input for all NN are a historical of signal error and one with actual set point, therefore, the first receives a tapped delay line with 19 historical of signal error and one with actual set point, the same form for the second and third [10, 11].

One NN is always added in the control system when the SD slowly to decreasing, the Table 1 shows these values in initial epoch of the train, one and two iteration after, one before penultimate, penultimate and last time of the train of NN. By Table 1 whenever a new NN is added, so the SD turn to decrease soon.

Table 1 MNFEL Automatic: Standard Deviation of the error of plant

Strategy / Time	Begin	Begin+1	Begin+2	End-2	End-1	End
FEL	0.076491	0.076237	0.076021	0.075095	0.075091	0.075093
MNFEL with 2 NN	0.075078	0.075072	0.075067	0.075008	0.075008	0.075008
MNFEL with 3 NN	0.075006	0.075004	0.075003	0.074984	0.074984	0.074984

5 Conclusion

FEL is an efficient collaborative strategy to be used with the CFC whether it is not desired to replace the CFC by non conventional strategies, but only improved it. However the MNFEL can improve so much the results obtained with FEL.

This work was devoted to present an innovation in the MNFEL strategy, an automatic insertion of neural networks. This mechanism of automatic insertion dispenses the specialist's knowledge and can be very necessary in many industrial process. Emphasis the use of SD not only say when add a new NN, but can help in the architecture of NN, because if a new NN enter in the control system and does not decrease the SD, can change the architecture and see the new result.

For further works, it is planned to extend the comparison, including other non conventional control strategies, as fuzzy control.

References

1. Almeida Neto, A.: Applications of Multiple Neural Networks in Mechatronic Systems. PhD thesis, Technological Institute of Aeronautics, BR (2003) (in Portuguese)
2. Almeida Neto, A., Goes, L.C.S., Nascimento Jr., C.L.: Multiple neural networks in flexible link control using feedback-error-learning. In: Proceedings of the 16th Brazilian Congress of Mechanical Engineering (2001)
3. Almeida Neto, A., Goes, L.C.S., Nascimento Jr., C.L.: Multi-layer feedback-error-learning for control of flexible link. In: Proceedings of the 2nd Thematic Congress of Dynamics, Control and Applications 2, pp. 2281–2289 (2003)
4. Gomi, H., Kawato, M.: Learning Control for a Closed Loop System Using Feedback-Error-Learning. In: Proceedings of the 29th Conference on Decision and Control, Honolulu, Hawaii, December 1990, vol. 6, pp. 3289–3294 (1990)
5. Haykin, S.: Neural Networks: A Comprehensive Foundation, 2nd edn. Prentice Hall, Englewood Cliffs (1999)
6. Kawato, M., Furukawa, K., Suzuki, R.: A hierarchical neural-network model for control and learning of voluntary movement. Biol. Cybernetics 57, 169–185 (1987)
7. Kurosawa, K., Futami, R., Watanabe, T., Hoshimiya, N.: Joint angle control by fes using a feedback error learning controller. IEEE Transactions on Neural Systems & Rehabilitation Engineering 13(3), 359–371 (2005)
8. Nakanishi, J., Schaal, S.: Feedback error learning and nonlinear adaptive control. Neural Networks 17(10), 1453–1465 (2004)
9. Ribeiro, P.R.A., Almeida Neto, A., Oliveira, A.C.M.: Using feedback-error-learning for industrial temperature control. In: CACS International Automatic Control Conference (2009)
10. Ribeiro, P.R.A., Almeida Neto, A., Oliveira, A.C.M.: Multi-network-feedback-error-learning in pelletizing plant control. In: 2nd IEEE International Conference on Advanced Computer Control - ICACC (2010)
11. Ribeiro, P.R.A., Costa, T.S., Barros, V.H., Almeida Neto, A., Oliveira, A.C.M.: Feedback-error-learning in pelletizing plant control. In: ENIA - 7th Brazilian Meeting on Artificial Intelligence (2009)
12. Ruan, X., Ding, M., Gong, D., Qiao, J.: On-line adaptive control for inverted pendulum balancing based on feedback-error-learning. Neurocomputing 70(4-6), 770–776 (2007)
13. Vale. Company, http://www.vale.com (Accessed in: January 10, 2010)
14. Yiwei, L., Shibo, X.: Neural network and pid hybrid adaptive control for horizontal control of shearer. In: ICARCV, pp. 671–674 (2002)

An Optimized 3D Surface Reconstruction Method Using Spatial Kalman Filtering of Projected Line Patterns

An-Qi Shen and Ping Jiang

Abstract. Real-time 3D surface reconstruction is a widely interested technique in the manufacturing industry. In this paper, an effective 3D surface reconstruction method using the active stereo with the projection of straight line patterns is proposed. For the purpose of real-time applications, 3D surface reconstruction is achieved by using two simple line projections. A depth-map is obtained by searching the projected lines. The final 3D surface is constructed by applying a spatial Kalman filter to the measured depth-map. Because it uses very simple line projections and searches the robust line patterns, it can be a fast, reliable and cost saving 3D measurement in comparison with the sinusoidal fringes projection methods.

1 Introduction

As the increasing demand in industrial automation and 3D animation, 3D shape measurement has become an active research area. In order to facilitate stereo matching, active stereo is often adopted by industrial 3D modeling systems, where a single camera and structured light projection system is used. The actively projected patterns, such as sinusoidal fringes projection and coded pixel projection, make the correspondence matching fast and reliable.

The sinusoidal fringes method with 4 phase shifted images can have reliable and accurate measurement [1, 2] but it relies on correct phase unwrapping, which becomes very difficult when discontinuous surface or occlusion is observed [9, 10]. Furthermore, the realization of a highly aligned and reliable phase shifted sinusoidal projections are hard and expensive. Significant effort has been made to

---

An-Qi Shen · Ping Jiang
School of computing, informatics and media, University of Bradford,
West Yorkshire, BD7 1DP, Bradford, United Kingdom
e-mail: {A.Shen,P.Jiang}@bradford.ac.uk

E. Corchado et al. (Eds.): SOCO 2010, AISC 73, pp. 179–186.
springerlink.com

achieve global unwrapping but all of them are with some restrictions [3, 4, 5]. The Gray code method and other binary codes methods [11, 12], even color code methods [13], are also commonly used for 3D surface reconstruction. Usually they have less accuracy and can be fallible for different test environments and surface materials [14].

This paper presents a 3D surface measurement application for quality control of car-bodies produced by pressing machines. The shape of the produced car-bodies can be changed due to the abrasion of the pressing tools. For the improvement of production quality of an automobile line, a fast and reliable 3D measurement system is developed. It takes two simple line projection images and detects only bright line patterns in the image. A spatial Kalman filter is used for 3D surface reconstruction based on the deformation of the line patterns. Section 2 presents the system structure and algorithms of the method. Section 3 gives the implementation and experimental results. Conclusion is given in the last section.

2 3D Measurement by Structured Light Projection

2.1 System Structure

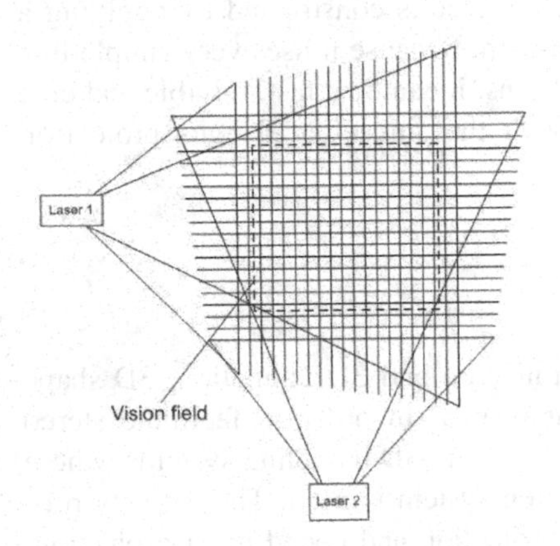

Fig. 2.1 System structure with two lasers mounted perpendicularly for projecting lines and an overhead camera for capturing images.

In order to measure 3D shape, horizontal and vertical lines are projected onto a reference plane Due to the tilt angle between the image plane and the reference plane, a rectangle in the image plane will be projected as a trapezoid due to the perspective projection. For simplification of image processing, we use two laser projectors mounted perpendicularly with their x directions in parallel with the reference plane. Each one projects a fixed number of horizontal lines. With an overhead camera parallel to the reference plane as shown in Fig. 2.1, square grids on the reference plane can be easily observed by the camera. This configuration has two advantages for image processing over the traditional sinusoidal fringes with a single projector. First detection of binary line patterns is more robust than the continuous phase detection; second, searching parallel lines along x or y direction can determine equal depth positions quickly.

2.2 Height Calculation

As we know, the straight lines projected onto a plane are still straight. But when projected onto an object with a certain height, they are deformed. Any deformation implies changes on height. Fig. 2.2 indicates how the height change can be calculated by the detection of the deformation of a light pattern.

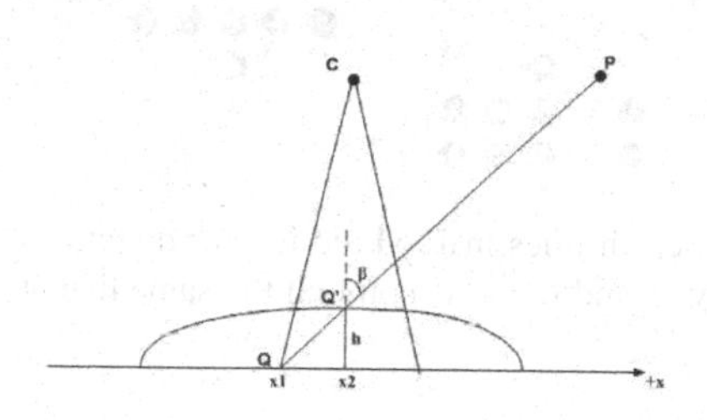

Fig. 2.2 Illustration of height detection

Point P denotes the exit pupil of the projector. Point C denotes the entrance pupil of the camera. Q is the original projected point without the object. Q' is the new projected point due to presence of the object. β is the angles of the projector with respect to the vertical line. A camera is mounted in parallel with the reference plane. Consider the camera is located far enough from the object that the angle between the camera light axis and the vertical line can be omitted.

The height $h_{i,j}$ at a point $Q_{i,j}$ can be calculated by the equation (2.1):

$$h_{i,j} = \frac{k_{i,j}}{\tan \beta_{i,j}} \cdot (x_2 - x_1) \tag{2.1}$$

where βi,j is a constant at each pixel; x_2, x_1 are the coordinates of the projected points perpendicular to the projected line; $k_{i,j}$ is the scale factor from pixel distance to real world distance, which can be obtained by calibration. Thus, the height $h_{i,j}$ can be determined by measuring the displacement in the image plane.

2.3 Searching Lines

Due the configuration in Fig. 2.1, the relative height along a line can be obtained by searching along the projected line and calculating displacements on x and y directions. A line searching algorithm is developed for the purpose:

Browse the whole image to find the intersection points and search lines along four directions using the search sequence shown in Fig. 2.3.

The height $h_{in,jn}$ at any point $Q_{in,jn}$ on a line can be calculated as:

$$h_{i_n,j_n} = \frac{k_{in,jn}}{\tan \beta_{in,jn}} \cdot \delta_n + h_{i_{n-1},j_{n-1}} \quad \delta_n = \begin{cases} i_n - i_{n-1} & horizontal\, line \\ j_n - j_{n-1} & vertical\, line \end{cases} \tag{2.2}$$

Loop until all intersection points are searched.

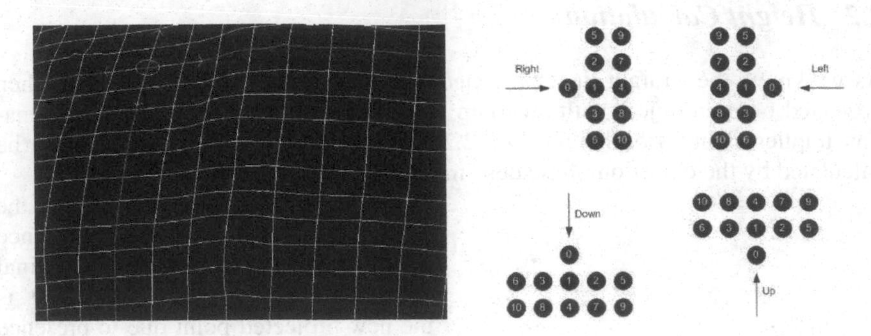

Fig. 2.3 Start point is marked with green circle. Search lines marked red in four directions from this point. Current point is marked as 0. Try to find the next point on the same line at position marked from 1 to 10.

2.4 Height Adjustment

The relative heights of points on each line differ from different reference heights. All the lines in a connected region should be adjusted to ensure that a common initial height is taken as a reference.

Find the first intersection point $Q_{i0,j0}$, search along vertical line UP and DOWN. When reach another intersection point, and then search along horizontal line LEFT and RIGHT to ensure the relative heights on the horizontal line is based on this intersection point on vertical line. Loop until all intersected horizontal lines are scanned.

After the adjustment, the height of any point which can be routed from the first point $Q_{i0,j0}$ is updated with a common reference height h_0 of $Q_{i0,j0}$. The collection of these points makes up a connection region in the image. At any point in the connected region, the height can be calculated as:

$$h_{i_n,j_n} = \sum_{m=0}^{n-1} \frac{k_{i_{m+1},j_{m+1}}}{\tan \beta_{i_{m+1},j_{m+1}}} \cdot \delta_{m+1} + h_0 \quad \delta_{m+1} = \begin{cases} i_{m+1} - i_m & horizontal line \\ j_{m+1} - j_m & vertical line \end{cases} \tag{2.3}$$

3 3D Surface Reconstruction by a Kalman Filter

Using the method described in the last section, we can get a rough height map for each region, which involves measurement noises, discrete errors from a camera and other uncertainties, if we assume all the noise obeys the Gaussian distribution. [6] Kalman filter can be used here to get a fine 3D measurement.

The Kalman filter is a set of mathematical equations that provides an efficient computational (recursive) means to estimate the state of a process, in a way that

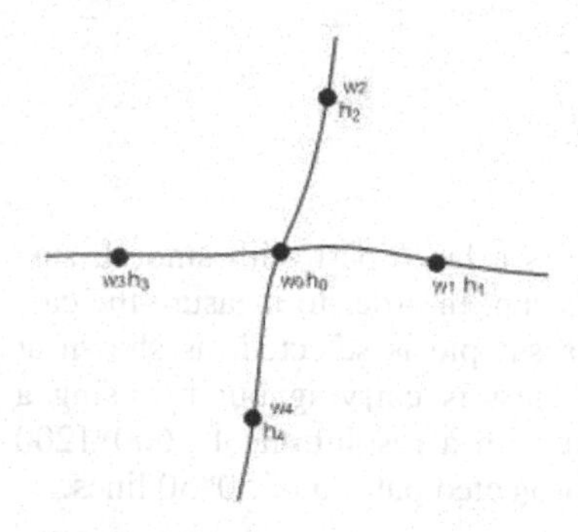

Fig. 3.1 For any intersection point Q_0 and its neighbors Q_1, Q_2, Q_3 *and* Q_4, they have height estimate h_i, i=0,..,4, and the associated uncertainty w_i.

minimizes the mean of the squared error [8]. In terms of the 3D measurement through projected lines, searching each line can give us a relative height estimate to a reference point. The intersection of two perpendicular lines means its height can be obtained by the combination of estimates from the four neighbors, which will be superior to any individual. In this paper, a two dimensional Kalman filter applying to the grids is developed for fusion of height estimates.

In order to develop a Kalman filter to have the optimal estimate of the height, the system model can be assumed to be $h_0(k+1)=h_0(k)$. The prior estimate from this model can be written as $\hat{h}_0^- = h_0(k)$, the prior variance is $P^-(k) = P(k)$.

According to equation (2.3), h_0 can be estimated from any of its neighbor h_i. We have its measurement model:

$$z_0 = h_i + K \cdot \delta_{0,i} + w_{i,0} \quad \delta_{0,i} = \begin{cases} y_0 - y_i & horizontal \quad line \\ x_0 - x_i & vertical \quad line \end{cases} \tag{3.1}$$

where $w_{i,0}$ with a distribution of $N(0,R)$ is the estimate variance from h_i to h_0. The Kalman gain can thus be calculated as

$$K(k) = P^-(k)(P^-(k) + R)^{-1} \tag{3.2}$$

The posterior estimate from h_i can be obtained as

$$\begin{aligned} &Posterior \quad Estimate: \quad \hat{h}_0 = \hat{h}_0^- + K(k)(z_0 - \hat{h}_0^-) \\ &Posterior \quad Variance: \quad P(k) = (1 - K(k))P^-(k) \end{aligned} \tag{3.3}$$

After a certain times of iterative calculation, the height will get into a dynamic balance state around the true state within a desired tolerance.

Therefore, equations (3.1)-(3.3) provide a set of equation for estimate of the height at each intersection point according to its neighbors. At the same time, this point is also the neighbor of its neighbors. So the adjustment can be spread from one point to the whole image map. Finally, it will reach a dynamic balance state around the original surface, which is an optimal surface based on all measurements.

4 Implementation

4.1 Sample Object

Fig. 4.1 Sample object

The car-body is a large part with smooth surface without steep. In order to measure the car-body depth, a sample is selected, as shown at Fig. 4.1. The test is carrying out by using a mono camera with a resolution of 1600*1200 pixels and a projected pattern of 50*50 lines.

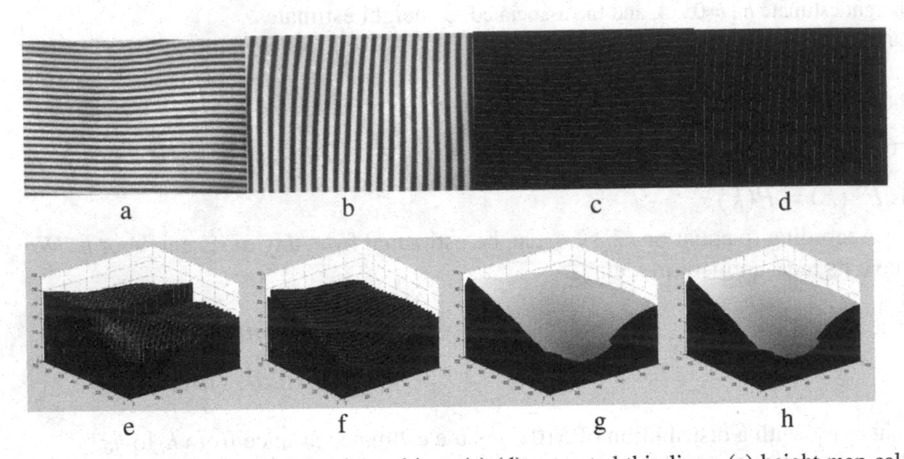

Fig. 4.2 (a) (b) line projection of the object. (c) (d) extracted thin lines. (e) height map calculated after searching lines. (f) height map after adjusting height (g) constructed 3D surface before applying Kalman filter. (h) constructed 3D surface after applying Kalman filter.

Table 4.1 Total variance after different iterate times. The variance is keeping decreasing significantly. According to an image with the size of 1600*1200 pixels, consider the acceptable tolerance is 200 which is relevant to 0.01% of the whole image.

Iterate times	1	2	3	4	5	6	7	8
Total variance	723.8	996.5	209.3	203.5	65.3	65.3	42.2	40.3

Table 4.2 Cycle time after applying 5 times Kalman filter iterate calculation in different size of image (unit: ms)

Image size	640*480	800*600	1024*768	1600*1200
Image processing time	2021	3283	4852	7420
Whole cycle time	4781	5961	7418	10142

At widely common used image size 640*480, the whole cycle time is less than 5s, which is half of the 4 phase shifted sinusoidal method. For the high accuracy, the whole cycle time is around 10s, while using 1600*1200 resolutions.

4.2 Other Samples

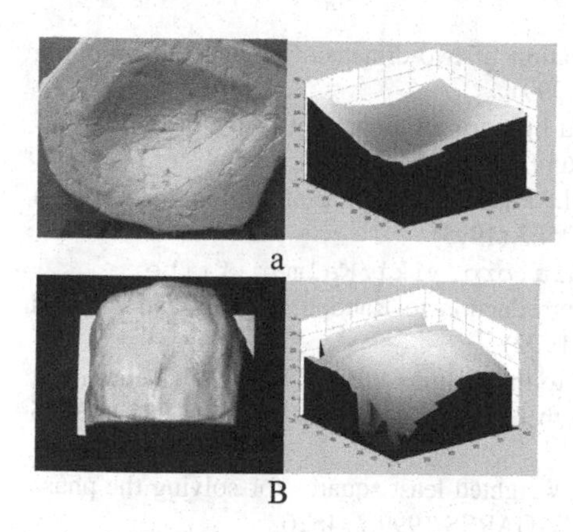

Fig. 4.3. Reconstructed 3D surface on different objects

Use the same system setup and try to do the test on different objects, which have variant surfaces type and materials. We can see, this method can get their surface as well.

Table 4.3 result of other objects in the image size of 1600*1200 after 5 times iterate calculation

Object	a	b
Total variance before iteration	867.3	653.4
Total variance after iteration	92.1	67.2
whole cycle time(ms)	10410	8882

5 Conclusion

This paper presented an active stereo system for real-time 3D measurement, where simple line patterns are projected. From the lessons learnt from the experimental study, the advantages and disadvantages can be summarized below:

Advantage

1. It's using simple and fixed light pattern which is easy established.
2. Light pattern can be achieved by laser. It's suitable for industry use for its high performance comparing with projector.
3. It is robust for most system setup. No homographic transform is applied.
4. Heights are adjusted in a small region. And the adjustment is only depended on its neighbors. No need to make global decision.
5. Fast to reconstruct a 3D surface, according to the whole cycle time.

Disadvantage

1. Clear lines image required. Diffuse surface materials are not suitable.
2. Isolated area will be independent to other areas due to lack of line connection.
3. The accuracy of the heights depends on the number of lines. Because it requires high signal-to-noise ratio, the density of lines should not be too high.
4. The image processing time is variable. For a complex object, it might take longer.

References

[1] Robinson, D.W., Reid, G.T.: Interferogram analysis: Digital Fringe Pattern Measurement Techniques. IOP Publishing Ltd. (1993)
[2] Wu, L.S., Peng, Q.J.: Research and development of fringe projection-based methods in 3D shape reconstruction. Journal of Zhejiang University, SCIENCE A (2006) doi:10.1631/jzus.2006.A1026
[3] Strand, J., Taxt, T.: Performance evaluation of two-dimensional phase unwrapping algorithms. Appl. Opt. 38, 4333–4343 (1999)
[4] Park, J., Kim, C., et al.: Using structured light for efficient depth edge detection. Image and Vision Computing 26, 1449–1465 (2008)
[5] Quiroga, J.A., Gonzalez-Cano, A., et al.: Phase-unwrapping algorithm based on an adaptive criterion. Appl. Opt. 34, 2560–2563 (1995)
[6] Kalman Filter, http://en.wikipedia.org/wiki/Kalman_filter
[7] Brown, R.G., Hwang, P.Y.C., et al.: Introduction to Random Signals and Applied Kalman Filtering, 2nd edn., New York (1992)
[8] Welch, G., Bishop, G.: An Introduction to the Kalman Filter, TR 95-041 Department of Computer Science University of North Carolina at Chapel Hill, NC 27599-3175 (1995)
[9] Lovergine, Stramaglia, F.P., et al.: Fast weighted least squares for solving the phase unwrapping problem (1999), doi:10.1109/IGARSS.1999.774626
[10] Hsu, Chen, R.L., et al.: Local weight selection for two-dimensional phase unwrapping (1999), doi:10.1109/ICIP.1999.817270
[11] Pagess, J., Salvi, J., et al.: Overview of coded light projection techniques for automatic 3D profiling. In: International Conference on Robotics and Automation (ICRA), pp. 133–138 (2003)
[12] Ferrari, M.B., et al.: A Coded modulation schemes based on partial Gray mapping and unpunctured high rate turbo codes (2005), doi:10.1049/ip-com:20050292
[13] Balas, C.: An imaging colorimeter for non contact tissue color mapping (1997), doi:10.1109/10.581936
[14] Wu, H.B., Chen, Y., et al.: 3D Measurement Technology by Structured Light Using Stripe-Edge-Based Gray Code. Journal of Physics: Conference Series 48, 537–541 (2006)

Decision Making and Quality-of-Information

Paulo Novais, Maria Salazar, Jorge Ribeiro, Cesar Analide, and José Neves

Abstract. In Group Decision Making based on argumentation, decisions are made considering the diverse points of view of the different partakers in order to decide which course of action a group should follow. However, knowledge and belief are normally incomplete, contradictory, or error sensitive, being desirable to use formal tools to deal with the problems that arise from the use of uncertain and even not precise information. On the other hand, qualitative models and qualitative reasoning have been around in Artificial Intelligence research for some time, in particular due the growing need to offer support in decision-making processes, a problem that in this work will be addressed in terms of an extension to the logic programming language and based on an evaluation of the Quality-of-Information (QoI) that stems out from those extended logic programs or theories. We present a computational model to address the problem of decision making, in terms of a multitude of scenarios, also defined as logic programs or theories, where the more appropriate ones stand for the higher QoIs values.

1 Introduction

Commonly, knowledge and belief are incomplete, contradictory, or error sensitive, being desirable to use formal tools to deal with these problems [1,2]. Logic and Logic programs [3] have emerged as attractive knowledge representation formalism and an approach to solving search problems. In the past few decades, many

Paulo Novais · Cesar Analide · José Neves
CCTC, Department of Informatics, University of Minho, Braga, Portugal
e-mail: {pjon,analide,jneves}@di.uminho.pt

Maria Salazar
Centro Hospitalar do Porto, EPE, Porto, Portugal
e-mail: msalazar@chporto.min-saude.pt

Jorge Ribeiro
School of Technology and Management, Viana do Castelo Polytechnic Institute,
Viana do Castelo, Portugal
e-mail: jribeiro@estg.ipvc.pt

E. Corchado et al. (Eds.): SOCO 2010, AISC 73, pp. 187–195.
springerlink.com

non-classical techniques for modeling the universe of discourse and reasoning procedures of intelligent systems have been proposed [4, 5, 6]. A part from the need to treat the problem of uncertain information there exists a second need related to the problem of incomplete information. Logic Programming presents a powerful and attractive knowledge representation and reasoning formalism to solve search problems in environments with defective information. For example, Hommersom and Colleagues [7] work is a good example of quality evaluation using logic. They used abduction and temporal logic for quality-checking of medical guidelines, proposing a method to diagnose potential problems in a timeline, regarding the fulfillment of general medical quality criteria at a meta-level characterization. They explored an approach which uses a relational translation to map the temporal logic formulas to first-order logic and a resolution-based theorem prover.

The objective is to build a quantification process of the Quality-of-Information (QoI) that stems from a logic program or theory during an evolutive process that aims to solve a problem in environments with incomplete information. It is presented a model for group decision making with quality evaluation, along with the several stages of the decision making process in the context of a Group Decision Support System (GDSS) for VirtualECare [9].

2 The Computational Model

With respect to the computational model it was considered an extension to the language of Logic Programming with two kinds of negation, classical negation, $\neg$, and default negation, `not`. Intuitively, `not p` is `true` whenever there is no reason to believe `p` (close world assumption), whereas $\neg p$ requires a proof of the negated literal. An Extended Logic Program (ELP for short) [10], on the other hand, is a finite collection of rules of the form [4]:

$$q \leftarrow p_1 \wedge \ldots \wedge p_m \wedge not\ p_{m+1} \wedge \ldots \wedge not\ p_{m+n}$$

$$?\ p_1 \wedge \ldots \wedge p_m \wedge not\ p_{m+1} \wedge \ldots \wedge not\ p_{m+n}$$

where $?$ is a domain atom denoting falsity, and q and every p_i are literals, i.e. formulas like a or $\neg a$, being a an atom, for $m, n \in N_0$. ELP introduces another kind of negation: strong negation, represented by the classical negation sign $\neg$. In most situations, it is useful to represent $\neg A$ as a literal, if it is possible to prove $\neg A$. In EPL, the expressions A and `not A`, being `A` a literal, are extended literals, while `A` or $\neg$`A` are simple literals.

Every program is associated with a set of abducibles. Abducibles can be seen as hypotheses that provide possible solutions or explanations of given queries, being given here in the form of exceptions to the extensions of the predicates that make the program. To reason about the body of knowledge presented in a particular program or theory, set on the base of the formalism referred to above, let us consider a procedure given in terms of the extension of a predicate called *demo*, using

ELP. This predicate allows to reason about the body of knowledge presented in a particular domain, set on the formalism referred to above. Given a question it returns a solution based on a set of assumptions. This meta predicate (demo) will be defined as: A meta theorem-solver for incomplete information represented by the signature *demo:T,V →{true,false}*, infers the valuation *V* of a theorem *T* in terms of the truth values *false* (or *0*), *true* (or *1*) and *unknown* (with truth values in the interval *]0,1[*), according to the following set of productions:

```
demo(T, true) ← T.
demo(T, false) ← ¬ T.
demo(T, unknown) ← not T, not ¬ T.
```

As a simple example, let us consider the following set of predicates, that stand for themselves:

```
itch: Name x Value
fever: Name x Value
pain: Name x Value
```

where the first argument denotes the name of the patient and the second one the truth value (or degree of confidence) that one has on the former. The extension of predicate *itch* may now be given in the form (program 1):

$\neg itch(X,Y) \leftarrow not\ itch(X,Y),\ not\ abducible_{itch}(X,Y).$

$abducible_{itch}(X,Y) \leftarrow itch(X,\bot).$

$itch(kevin,\bot).$

$itch(john,1).$

$abducible_{itch}(carol,0.6).$

$abducible_{itch}(carol,0.8).$

$?\ ((abducible_{itch}(X_1,Y_1) \lor abducible_{itch}(X_2,Y_2)) \land \neg (abducible_{itch}(X_1,Y_1) \land abducible_{itch}(X_2,Y_2)))$

Program 1. Extension of the predicate *itch*

where the first clause denotes the closure of the predicate *itch*. In the second clause the symbol '⊥' stands for a null value, in the sense that it subsumes that *Y* may take any truth value in the interval *[0,1]*. The fourth clause denotes that the truth value of *itch* for the patient *john* is *1*. The clauses five and six denote the fact that the truth value of *itch* for patient *carol* is either *0.6* or *0.8*, or even both. The seventh clause stands for the invariant that implements the *XOR* operator, i.e. it states that the truth value of *itch* for the patient *carol* is either *0.6* or *0.8*, but not both.

The extension of predicate `fever` may now be given in the form (program 2):

```
¬fever(X,Y) ← not fever(X,Y), not abducible_fever(X,Y).

abducible_fever(X,Y) ←fever(X,⊥).

fever(carol,⊥).

fever(kevin,1).

abducible_fever(john,0.50).

abducible_fever(john,0.75).

?((abducible_fever(X1,Y1) ∨ abducible_fever(X2,Y2)) ∧ ¬
(abducible_fever(X1,Y1) ∧ abducible_fever(X2,Y2)))
```

Program 2. Extension of the predicate *fever*

The extension of predicate `pain` may now be given in the form (program 3):

```
¬pain(X,Y) ← not pain(X,Y), not abducible_pain(X,Y).

abducible_pain(X,Y)←pain(X, ⊥).

pain(carol, ⊥).

pain(kevin, 1).

abducible_pain(john,0.3).

abducible_pain(john,0.45).

abducible_pain(john,0.57).
```

Program 3. Extension of the predicate *pain*

In program 3 the last three clauses denote the case where the truth value for `pain` for patient `john` is unknown, although in the set $\{0.3, 0.45, 0.57\}$.

3 Quality-of-Information

In decision making processes [9, 11] it is necessary to search only the most promising search paths. Each path must be tested on their ability to adapt to changing environments, to make deductions and draw inferences, and to choose the most appropriate course of action from a wide range of alternatives. The optimal path in an ELP context is the logic program or theory that models the universe of discourse and maximizes its Quality-of-Information (QoI) factor. Let i $(i \in [1,m])$ represent the predicates whose extensions make an extended logic program that models the universe of discourse, as it is given above in terms of the predicates `itch`, `fever`, and `pain`, where j $(j \in [1,n])$ denote the attributes of those predicates. Let $x_j \in [min_j, max_j]$ be a value for attribute j. To each predicate is also associated a scoring function `Vij[minj, maxj] → [0,1]`, that gives the score predicate `i` assigned to a value of attribute `j` in the range of its acceptable values, i.e. its domain (for simplicity, scores are kept in the interval `[0,1]`).

The `Quality-of-Information` (`QoI`) with respect to a generic predicate `P` can be analyzed in four situations and can be measure in the interval `[0,1]`, when the information is positive and negative, when the information is unknown but can be selected from one or more values, and when the information is unknown but can be derived from a set of values, but only one can be selected. If the information is known (positive) or false (negative) the (`QoI`) for the predicate term under consideration is `1`. For situations where the value is *unknown* the `QoI` is given by:

$$QoI_P = \lim_{N \to \infty} \frac{1}{N} = 0 (N >> 0)$$

For situations when the information is unknown but can be derived from a set of values, $QoI_P = 1/Card$, where *Card* denotes the cardinality of the exception set for `P`, if the exception set is disjoint. If the exception set is not disjoint, the quality-of-information is given by:

$$QoI_P = \frac{1}{C_1^{Card} + \cdots + C_{Card}^{Card}}$$

where C_{Card}^{Card} is a card-combination subset, with *Card* elements. The next element of the model to be considered is the *relative importance* that a predicate assigns to each of its attributes under observation, i.e. w_{ij} stands for the relevance of attribute j for predicate i. It is also assumed that the weights of all predicates are normalized, that is:

$$\forall i \sum_{j=1}^{n} w_{ij} = 1 \text{, for all } i.$$

On the another hand, the predicate scoring function, when associated to a value $x=(x_1, \ldots, x_n)$ in a multi-dimensional space, it is defined in terms of its attribute domains in the form:

$$V^i(x) = \sum_{j=1}^{n} w^i{}_j * V^i{}_j(x_j)$$

Therefore, it is viable to measure the *QoI* that occurs as a result of invoking a logic program to prove a theorem (e.g. Theorem), by posting the $V_i(x)$ values into a multi-dimensional space and projecting it onto a two dimensional one. For example, for patient john, a logic program or theory P may be got in terms of the logic programs 1, 2 and 3 referred to above, being depicted in the form:

```
¬itch(X,Y) ← not itch(X,Y), not abducible_itch(X,Y).
itch(john,1).
¬fever(X,Y) ← not fever(X,Y), not abducible_fever(X,Y).
abducible_fever(john,0.50).
```

```
abducible_fever(john,0.75).
?((abducible_fever(X1,Y1) ∨ abducible_fever(X2,Y2)) ∧ ¬
(abducible_fever(X1,Y1) ∧ abducible_fever(X2,Y2)))

¬pain(X,Y) ← not pain(X,Y), not abducible_pain(X,Y).
abducible_pain(john,0.3).
abducible_pain(john,0.45).
abducible_pain(john,0.57).
```

whose *QoI* is presented in Figure 1

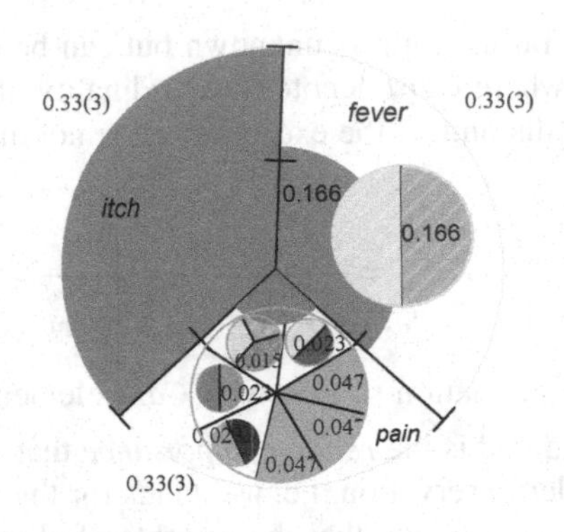

Fig. 1 A measure of the QoI for the Logic Program or theory *P* referred to above

4 Decision Making in VirtualECare

The VirtualECare project embodies an intelligent multi-agent system aimed to monitor and interact with its users, targeted to elderly people and/or their relatives. The system is designed to have several services, beyond the health related ones. It will be connected not only to healthcare institutions, but also with user's relatives, leisure centers, training facilities and shops, just to name a few [9].

The VirtualECare GDSS is a knowledge-driven Decision Support Systems (DSS) [8], that relies on a database (or knowledge base), and models representations of the world, following a proof-theoretical approach to computing, that addresses the truth value of a theorem to be proven in terms of the *QoI* of the terms that make the extension of a predicate or predicates under invocation [10].

Our approach of a VirtualECare GDSS follows Simon's empirical rationality [12]. The *Intelligence* stage occurs continuously, as the GDSS interacts with other components of the VirtualECare system. Identified problems that call for an action triggers the formation of a group decision. This group formation is conducted in the pre-meeting phase, when a facilitator must choose the partakers. In order to form the "best" group one must evaluate the *QoI* on hand of possible participants, and not about the participants themselves, registered in the knowledge base system. The *Design* and *Choice* phases occur in the in-meeting stage. In the In-Meeting phase, the participants will be working in order to accomplish the meeting goals and to take de finest decisions. In order to accomplish this goal, the participants use a knowledge database and exchange information. Once again, the system must provide a measure of the *QoI* available. In the Post-Meeting phase it is important to evaluate the results achieved so far by the group, as well as how much each group member is acquit with the achieved results (satisfied/unsatisfied).

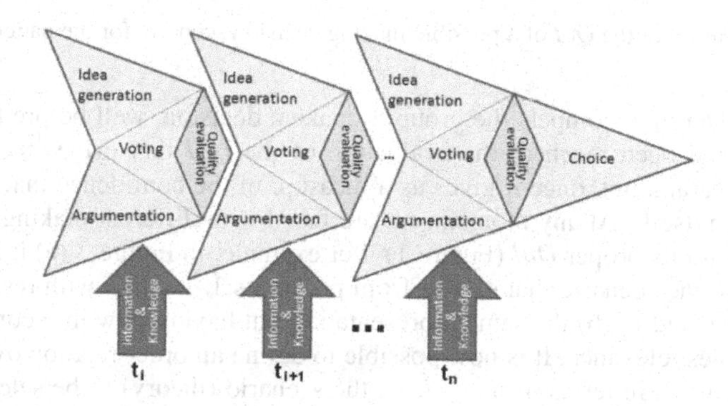

Fig. 2. In-meeting stage with several iterations

The in-meeting stage cycles through a set of iterations, similarly to the *circular logic of choice* of Nappelbaum [13]. In Nappelbaum model a sharpening spiral of the description of the problem cycles through option descriptions, value judgments and instrumental instructions, towards a prescribed choice. We further extend this approach, in line with Jones and Humphreys model of the Decision Hedgehog [14]. Instead of constructing and prescribing the solution to the decision problem within a procedural context of a single decision path, we suggest the exploration of potential different pathways to develop contextual knowledge, enabling collaborative authoring of outcomes.

In this way, the *QoI* is evaluated within each iteration, for every possible pathway. The knowledge system is scanned for the needed information with a previously agreed threshold of the *QoI* being measured [10, 15]. If the *QoI* measure does not reach the necessary threshold, new information and/or knowledge is searched for and the process restarts. Figure 2 illustrates the situation where the quality threshold is only reached on a step-by-step process, attaining a point in time when the decision is made. In each iteration, we can use different approaches to generate alternatives and criteria, namely Idea Generation, Argumentation and Voting techniques, to support the decision-making process [8].

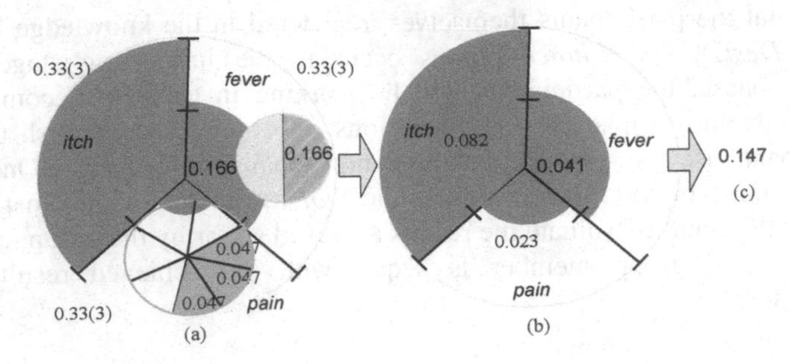

Fig. 3 A measure of the *QoI* of a possible making decision scenario for the patient *john*

Even when time compels the group to make a decision, well before the quality threshold had been reached, the evaluation of the *QoI* that drives the inference process is paramount, once it gives us a measure of the confidence that we put on the decision itself. At any moment, we are faced with different making scenarios, each one with its proper *QoI* (Figure 1). For example, in Figure 3 (a) it is depicted the *QoI* for one scenario that evolve from programs 1, 2 and 3 with respect to the patient *john*, and in (b) the same representation but having now into consideration the predicates relevance. It is now possible to define an order relation over the different scenarios, in terms of its *QoI*; i.e the scenario (theory) to be selected is the one that presents the greatest truth value.

5 Conclusions

Qualitative models and qualitative reasoning have been around in Artificial Intelligence research for some time, in particular due the growing need to offer support in decision-making processes. This area brings together research and evaluation projects in which healthcare decision-making plays a vital role. Indeed, decision-making with healthcare implications spans a broad area, and is relevant at the national, regional, local, and patient levels, and in the public and private spheres. The main focus of our studies in this field is decision-making at the patient level; although, as future work, we intend to study the problem at the organisational and societal levels. Our work addressed the problem of group decision making, modeling it in terms of a multitude of scenarios, defined as logic programs or theories, being its selection based on its soundness, here measured in terms of their QoIs values.

References

1. Kowalski, R.: The logical way to be artificially intelligent. In: Toni, F., Torroni, O. (eds.) Proceedings of the CLIMA VI. LNCS (LNAI), pp. 1–22. Springer, Heidelberg (2006)
2. Sheridan, F.: A Survey of techniques for inference under uncertainty. Artificial Intelligent Review 5(1), 89 (1991)

3. Ginsberg, M.L.: Readings in Nonmonotonic Reasoning, Los Altos, Califórnia, EUA. Morgan Kauffman Publishers, Inc., San Francisco (1991)
4. Neves, J.: A Logic Interpreter to Handle Time and Negation in Logic Data Bases. In: Proceedings of ACM 1984, The Fifth Generation Challenge, pp. 50–54 (1984)
5. Subrahmanian, V.: Probabilistic databases and logic programming. In: Proc. of the 17th International Conference of Logic Programming (2001)
6. Kakas, A., Kowalski, R., Toni, F.: The role of abduction in logic programming. Handbook of Logic in Artificial Intelligence and Logic Programming 5, 235–324 (1998)
7. Hommersom, A., Lucas, P.J.F., van Bommel, P.: Checking the quality of clinical guidelines using automated reasoning tools. Theory and Practice of Logic Programming 8(5-6), 611–641 (2008)
8. Lima, L., Novais, P., Bulas Cruz, J.: A Process Model For Group Decision Making With Quality Evaluation. In: Omatu, S., Rocha, M.P., Bravo, J., Fernández, F., Corchado, E., Bustillo, A., Corchado, J.M. (eds.) IWANN 2009. LNCS, vol. 5518, pp. 566–573. Springer, Heidelberg (2009)
9. Novais, P., Costa, R., Carneiro, D., Machado, J., Lima, L., Neves, J.: Group Support in Collaborative Networks Organizations for Ambient Assisted Living. In: Oya, M., Uda, R., Yasunobu, C. (eds.) Towards Sustainable Society on Ubiquitous Networks. IFIP, pp. 353–362. Springer, Heidelberg (2008)
10. Analide, C., Novais, P., Machado, M., Neves, J.: Quality of Knowledge in Virtual Entities. In: Encyclopedia of Communities of Practice in Information and Knowledge Management, pp. 436–442. Idea Group Inc., USA (2006)
11. Lu, J., Zhang, G., Ruan, D., Wu, F.: Multi-objective Group Decision Making: Methods, Software and Applications with Fuzzy Set Techniques. Electric and Computer Engineering. Imperial College Press,
12. Simon, H.A.: Models of Bounded Rationality: Empirically Grounded Economic Reason, vol. 3. MIT Press, Cambridge (1982)
13. Nappelbaum, E.: Systems logic for problem formulation and choice. In: Humphreys, P., et al. (eds.) IFIP TC8 WG8.3, Springer, Heidelberg (1997)
14. Jones, G., Humphreys, P.: The Decision Hedgehog: Enhancing Contextual Knowledge for Group Decision Authoring and Communication Support. In: Fifth International and Interdisciplinary Conference on Modeling and Using Context, CEUR-WS, Paris, France (2005)
15. Machado, J., Abelha, A., Novais, P., Neves, J., Neves, J.: Quality of service in healthcare units. Int. J. Computer Aided Engineering and Technology 2(4), 436–449 (2010)

The Gene Expression Programming Applied to Demand Forecast

Evandro Bittencourt, Sidney Schossland, Raul Landmann,
Dênio Murilo de Aguiar, and Adilson Gomes De Oliveira

Abstract. This paper examines the use of artificial intelligence (in particular the aplication of Gene Expression Programming, GEP) to demand forecasting. In the world of production management, many data that are produced in function of the of economic activity characteristics in which they belong, may suffer, for example, significant impacts of seasonal behaviors, making the prediction of future conditions difficult by means of methods commonly used. The GEP is an evolution of Genetic Programming, which is part of the Genetic Algorithms. GEP seeks for mathematical functions, adjusting to a given set of solutions using a type of genetic heuristics from a population of random functions. In order to compare the GEP, we have used the others quantitatives method. Thus, from a data set of about demand of consumption of twelve products line metal fittings, we have compared the forecast data.

1 Introduction

Vollman et al [8] pointed out that the competitive environment is having continuous changes, ranging from the technological planning to the strategic planning. According to [6], the concurrency nature is moving from the classical battle among companies to a supply chain contest. The supply chain is basically a set of installations connected by transportation routes.

In this scenario, the Supply Chain Management (SCM), a concept created by Keith Oliver from Booz Allen Hamilton in 1982, is considered like the last mine to be explored in order to increase the organizations competitive potentials.

Evandro Bittencourt · Sidney Schossland · Raul Landmann · Dênio Murilo de Aguiar · Adilson Gomes De Oliveira
UNIVILLE, Department of Management, Joinville, SC, Brazil
e-mail: evandrobitt@gmail.com, {sidney.schossland, raul.landmann,denio.murilo,adilson.gomes}@univille.br

E. Corchado et al. (Eds.): SOCO 2010, AISC 73, pp. 197–200.
springerlink.com

Thus, the keywords for the SCM's effectiveness are 'integration' and 'collaboration'. This collaborative planning is, in fact, a process in which the manufacturing and trade share information in the point of sales in order to prepare the joint estimate of demand [2].

The forecast establishment, therefore, is one of the most important activities of business management, and among them, the demand forecast is the main one, because, as states [7], is essential for strategic planning of sales, production and finances of any business.

2 Demand Forecasting's Techniques

There are various techniques for forecasting, which can be divided into two groups according to [7]: qualitative and quantitative. The qualitative techniques are based on people's opinion and judgment. The quantitative techniques use the data passed in mathematical models to project future demand, and is also subdivided into two groups: those based on time series and based on correlations.

3 Gene Expression Programming (GEP)

The genetic algorithm is an optimization process of complex functions which uses the Darwinian selection process for the evolution and definition of the best results. The genetic algorithm was originally proposed by J. Holland in the 70's and published by his students [3]. On the other hand, the genetic programming is an application of the genetic algorithm, which aims to optimize not values, but mathematical expressions.

Ferreira [4], proposed a new methodology, also known as the Gene Expression Programming (GEP), which is, in fact, an evolution of genetic programming. The word (chromosome) that defines a gene expression in GEP is divided into head, which contains operators and variables; and tail, containing only variables.

The chromosomes can have a gene or more than one gene, and the result of the generated expression is obtained by summing the expressions of each gene or other pre-defined operation. The GEP defines two languages, one for the genetic word called Karva, and another that defines a mathematical expression through an expression tree.

The translation is carried out by defining an expression tree from left to right in the various hierarchical levels according to the description of the genetic word. The length of the chromosome depends on the tails' calculation (t) according to the definition of head's length (h) and the maximum number of operators (n) (Equation 1).

$$t = h(n-1) + 1 \tag{1}$$

In GEP is also possible to create mathematical expressions with numeric constants [4]. Thus, besides the mathematical operators, we can use an extra symbol, the interrogation mark (?), in the operators set, which is related to an extension of the chromosome containing the index of the used constant, called the Dc, in a group of 10 constants which are formed randomly. The constants set also goes through a mutation development during the evolution process.

The GEP heuristic follows a process similar to genetic algorithm for the chromosomes evolution that form the initial population.

3.1 Creation of Initial Population and Testing

Before the creation of initial population we define the operators set, the size of the head, and the test values set. The process of creating the initial population's chromosomes happens in a random way. It is possible to form a chromosome with one or more genes. In this case, the result of the expression is made from the sum of the expressions represented by different genes or using another form of union, multiplication or division of gene expressions. After creating the initial population, each individual is expressed and executed from a given data set. With this result, the fitness of each individual can be calculated.

The stopping test can be based on the maximum fitness value or in another value, or even in a number of iterations (e.g. 1000 generations).

3.2 Selection, Elitism and Mutation

Each generation keeps up the best program, which not goes through the reproduction process. For the reproduction, we use a form of selection that takes into account the fitness value, or in other words, the program that has best fitness, has more chance to serve as genetic material for the next generation.

In this work we will use the roulette method, where the fitness of each program defines the segment of a roulette wheel, which can be larger or smaller. This segment defines a given program after a random spin.

We can make changes at any chromosome point. Both in the head and the tail, we should keep the condition of not using operations in the tail. The mutation rate varies from 5 % to 20 %, where we must select, for each chromosome, a value between 1 to 3 mutation points [4].

4 Proposed Process and Initial Results

The proposed method is to compare the forecast demand through the GEP and other quantitative techniques of twelve product line of metal fittings. An initial result for one of these products is presented in Table 1, the GEP is the best method considering the sum of the error. The future demand was expected by GEP and other techniques for trend based on data from the historical trend.

Table 1 Simulated demand (13-24 period) utilizing GEP and others methods

HD[a]		FD[b]		GEP		LD[c]		GT[d]		ET[e]		PT[f]		SI[g]	
p[h]	v[i]	p	v	v	e[j]	v	e	v	e	v	e	v	e	v	e
1	23159	13	30704	25584	0.17	25352	0.17	24153	0.21	24086	0.22	24879	0.19	26586	0.13
2	27863	14	26236	25811	0.02	25516	0.03	24171	0.08	24098	0.08	25026	0.05	29399	0.12
3	19562	15	34881	26049	0.25	25680	0.26	24187	0.31	24109	0.31	25174	0.28	25538	0.27
4	27494	16	25746	26348	0.02	25845	0.00	24202	0.06	24119	0.06	25323	0.02	28079	0.09
5	24935	17	35026	26870	0.23	26009	0.26	24216	0.31	24128	0.31	25473	0.27	28973	0.17
6	27686	18	24954	29667	0.19	26173	0.05	24229	0.03	24137	0.03	25624	0.03	32158	0.29
7	18756	19	26266	25128	0.04	26337	0.00	24242	0.08	24146	0.08	25775	0.02	28031	0.07
8	17389	20	20542	19354	0.06	26501	0.29	24254	0.18	24154	0.18	25928	0.26	30919	0.51
9	23660	21	31108	25169	0.19	26665	0.14	24265	0.22	24161	0.22	26081	0.16	31995	0.03
10	22685	22	25342	25625	0.01	26830	0.06	24276	0.04	24169	0.05	26236	0.04	35606	0.41
11	29604	23	26508	25850	0.02	26994	0.02	24286	0.08	24176	0.09	26391	0.00	31111	0.17
12	28628	24	28835	26000	0.10	27158	0.06	24296	0.16	24182	0.16	26547	0.08	34390	0.19
				Σ	1.31	Σ	1.35	Σ	1.76	Σ	1.79	Σ	1.39	Σ	2.45

[a] Historic Demand; [b] Future Demand ; [c] Linear Trend; [d] Geometric Trend ; [e] Exponential Trend ; [f] Potential Trend ; [g] Seazonal Indices ; [h] period ; [i] value ; [j] error.

References

1. Arnold, J.R.T.: Administracão de materiais, Atlas, São Paulo (1999)
2. Bertaglia, P.R.: Logística e gerenciamento da cadeia de abastecimento, Saraiva, São Paulo (2003)
3. Dejong, K.A.: An analysis of the behavior of a class of genetic adaptive systems. PhD Dissertation. Dept. of Computer and Communication Sciences, Univ. de Michigan, Ann Arbor (1975)
4. Ferreira, C.: Gene Expression Programming: A New Adaptive Algorithm for Solving Problems. Complex Systems 13, 87–129 (2001)
5. Haupt, R.L., Haupt, S.E.: Practical Genetic Algorithm. John Wiley & Sons, New York (1998)
6. Taylor, D.A.: Logística na cadeia de suprimentos. Pearson/Addison-Wesley, São Paulo (2005)
7. Tubino, D.F.: Planejamento e controle da producão - teoria e prática, Atlas, São Paulo (2007)
8. Vollman, T.E., Berry, W.L., Whybark, D.C., Jacobs, R.F.: Sistemas de planejamento e controle da producao para o gerenciamento da cadeia de suprimentos. Bookman, Porto Alegre (2006)

Brain Magnetic Resonance Spectroscopy Classifiers

Susana Oliveira, Jaime Rocha, and Victor Alves

Abstract. During the last decade, the Magnetic Resonance Spectroscopy modality has become an integrant part of the diagnostic routine. However, the visual interpretation of these spectra is difficult and few clinicians are trained to use the technique. In this study, sixty-eight spectra obtained from twenty-two multi-voxel spectroscopies were classified using three well-known classification algorithms: K-Nearest Neighbors (KNN), Decision Trees and Naïve Bayes. The best results were obtained using NaïveBayes that presented an average balanced accuracy rate around 75%, although K-Nearest Neighbors presented very good results in some situations. The obtained results lead us to conclude that it is possible to classify magnetic resonance spectra with data mining techniques for further integration in a Clinical Decision Support System which may help in the diagnosis of new cases.

1 Introduction

Magnetic Resonance Imaging (MRI) is a medical imaging modality developed in the early 1970s. It was realized that magnetic field gradients could be used to localize the magnetic resonance signal and to generate images that display magnetic properties, reflecting clinically relevant information [1]. Magnetic Resonance Spectroscopy (MRS) emerged from the physical discoveries of MRI as a new clinical tool in the 1990s. While MRI provides structural information, MRS provides qualitative information of a number of metabolites within the brain, and can also provide quantitative information if a reference of known concentration is used. These metabolites reflect aspects of neuronal integrity, cell membrane proliferation or degradation, energy metabolism and necrotic transformation of brain or tumor tissue. Hence, MRS is a non-invasive technique that can help in the diagnosis of different

Susana Oliveira · Victor Alves
Department of Informatics, University of Minho, Braga, Portugal
e-mail: susanaoliveira08@gmail.com, valves@di.uminho.pt

Jaime Rocha
Department of Neuroradiology, Hospital de Braga, Portugal
e-mail: jaimeroc@gmail.com

E. Corchado et al. (Eds.): SOCO 2010, AISC 73, pp. 201–208.
springerlink.com

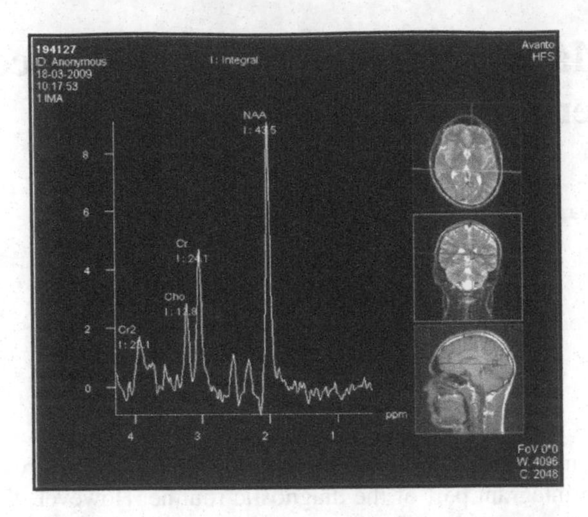

Fig. 1 1H-Magnetic resonance spectrum appearing grey matter of a normal volunteer

diseases, add information to imaging diagnosis and determine the tissue biochemical composition (metabolic profile) of a tumor. The spectrum has the general appearance of a plot of peaks along the x-axis, with the peak position depending on the resonant frequency of the associated metabolite (Fig. 1) [2].

During the last decade, the Magnetic Resonance Spectroscopy modality has become an integrant part of the diagnostic routine [3]. However, the visual interpretation of these spectra is difficult and few clinicians are trained to use the technique [4]. When analyzing recent developments, it becomes clear that the trend is to develop new methods for computer decision-making in clinical practice [5]. Thus, the use of artificial intelligence tools has become widely accepted in medical applications to support patient diagnosis more effectively, especially in, the application classification systems [6].

Recent publications have described the use of classification methods to analyze magnetic resonance spectra. Bezabeh, et al. (2002) analyzed "the potential use of proton magnetic resonance spectroscopy in combination with an appropriate statistical classification strategy in differentiating normal mesenchymal tissue from soft tissue sarcoma" and obtained good results. The classification was performed using Linear Discriminant Analysis (LDA) with the Leave-One-Out method. The statistical classification strategy gave much better results than the conventional analysis [7]. Siddall, et al. (2006) used a pattern recognition method to discriminate accurately subjects with low back pain from control subjects based on MRS. They concluded that MRS combined with an appropriate pattern recognition approach, was able to detect brain biochemical changes associated with chronic pain with a high degree of accuracy [8]. Luts, et al. (2007) studied the use of automated pattern recognition methods on magnetic resonance data with the goal to assist clinicians in the diagnosis of brain tumors. They used a multiclass classification system to assess the type and grade of a tumor. Moreover, they compared Least Squares Support Vector Machines (LS-SVMs) with LDA. They concluded that binary LS-SVMs can be extended to a multiclass classifier system and that this classifier system can be of great help in the diagnosis of brain tumors [9]. Finally, García-Gómez, et al. (2009) predicted models of automatic brain tumor classification using cases obtained in different centers. They concluded that the prediction of the tumor type of MRS is possible using classifiers developed from acquired data, in different hospitals with different instruments under the same acquisition protocols [10].

In this study various feature extraction and classification algorithms will be assessed. Our goal is to demonstrate that it is possible to classify magnetic resonance spectra with data mining techniques in order to use them in a Clinical Decision Support System (CDSS). This system will have the potential to help the physician in the diagnosis of new cases.

2 Methods

The spectra used in this study to obtain the classifiers were extracted from studies of patients previously diagnosed by a neuroradiologist. These studies were scanned by a “Siemens Magnetom Avanto 1.5T Magnetic Ressonance” at *Hospital de Braga,* following the same acquisition protocol (csi_se_135). This protocol was chosen since it is adequate for highly localized spectroscopy of the human brain, showing the main metabolites [11]. For each studied volunteer, two, three or four voxels with the same or a different pathology were used. The pathology was labeled as tumor, vascular, healthy and other. From the 22 multi-voxel spectroscopies used, we obtained 68 spectra (voxels). Of those, 36.8% were tumors, 20.6% vascular, 25.0% healthy and 17.6% other.

For each voxel, the following data was registered: patient number id, age, gender, race, metabolite of reference (NAA), position (in ppm), integral and the ratio in relation to the NAA for N-acetyl aspartate (NAA), Creatine (Cr), Choline (Cho) and Creatine2 (Cr2) metabolites. Table 1 presents the minimum, maximum and mean values by pathology for each metabolite.

Table 1 Minimum, maximum and mean by pathology and in general for the metabolites

Pathology	Values	NAA	Cr	Cho	Cr2
Tumor	Min./Max.	8.56/35.99	5.84/33.96	5.94/47.23	0.00/33.98
	Mean	19.63	18.50	25.12	13.13
Vascular	Min./Max.	13.84/41.90	15.54/30.54	17.21/30.87	7.91/21.56
	Mean	32.51	24.03	23.96	14.35
Healthy	Min./Max.	22.78/60.96	8.65/37.35	6.92/32.05	5.73/29.16
	Mean	38.98	23.27	23.32	16.75
Other	Min./Max.	17.48/40.45	13.40/38.47	14.94/49.40	13.94/42.61
	Mean	28.25	22.59	31.21	23.68

NAA is considered a neuronal marker. As most brain tumors are of non neuronal origin, NAA is absent or greatly reduced. Compared with the normal brain, Cr is generally reduced in astrocytomas and is almost absent in meningiomas, schwannomas and metastases. Cho reflects the metabolism of cell membranes [12].

Two data selection techniques were applied on the NAA, Cr, Cho and Cr2 integral. Feature Selection (FS) was used to select relevant attributes since that the data set could contain irrelevant attributes to the mining task and the adding of irrelevant or distracting attributes often "confuses" data-mining algorithms [12]. Principal Component Analysis (PCA) was used to transform the original values onto a new set of uncorrelated features. These new features are linear combinations of the original variables and they are ranked according to the variance they retain. Often the original data can be explained by a limited number of components [13]. The spectra were classified using three well-known classification algorithms: K-Nearest Neighbors (KNN), Decision Tree and Naïve Bayes. KNN classifies an unclassified sample based on the nearest k of a set of previously classified samples. The training samples are mapped into multidimensional feature space. This space is partitioned into regions by class labels. A point in the space is assigned to the class if it is the most frequent class label among the nearest training samples. Usually the L2-norm or Euclidean distance is used but other distances like Manhattan distance can also be used [14] [15]. Decision tree algorithms were originally intended for classification, but are powerful and popular for both classification and prediction. The attractiveness of tree-based methods is due largely to the fact that decision trees represent rules that humans can easily understand [16]. A decision tree consists of nodes where attributes are tested. The outgoing branches of a node correspond to all the possible outcomes of the test. At each node, the algorithm chooses the "best" attribute to partition the data into individual classes [11][17]. Naïve Bayesian classification is the popular name for a probabilistic classification based on the Bayesian Theorem [17] [18].

The models were evaluated using k-fold cross-validation, the initial data was randomly partitioned into k mutually exclusive subsets or "folds" each of them with approximately equal size. One of the k subsets is used as the test set and the others k-1 subsets are used as training sets. This process is repeated k times until all k subsets become the test set. Subsequently, the average of a performance measure (e.g. accuracy) across all k trials is computed [19] [20]. For the evaluation of the classifiers, a Balanced Accuracy Rate (BAR) was used. In a binary classifier A vs. B, the BAR is the average of the accuracy rate on the A and B classes (i.e. $\frac{1}{2}\left(\frac{a_A}{n_A}+\frac{a_B}{n_B}\right)$ where n_A and n_B denote the number of cases of the two classes and a_A and a_B the number of well-classified cases. BAR is especially useful when two classes are imbalanced, i.e. when one of the classes is more representative than the other.

3 Results and Discussion

All classification models were evaluated using 5-fold cross-validation [20]. In KNN, k was varied between 1 and the number of samples of the smaller class divided by the number of blocks. Euclidean metric was applied to calculate the

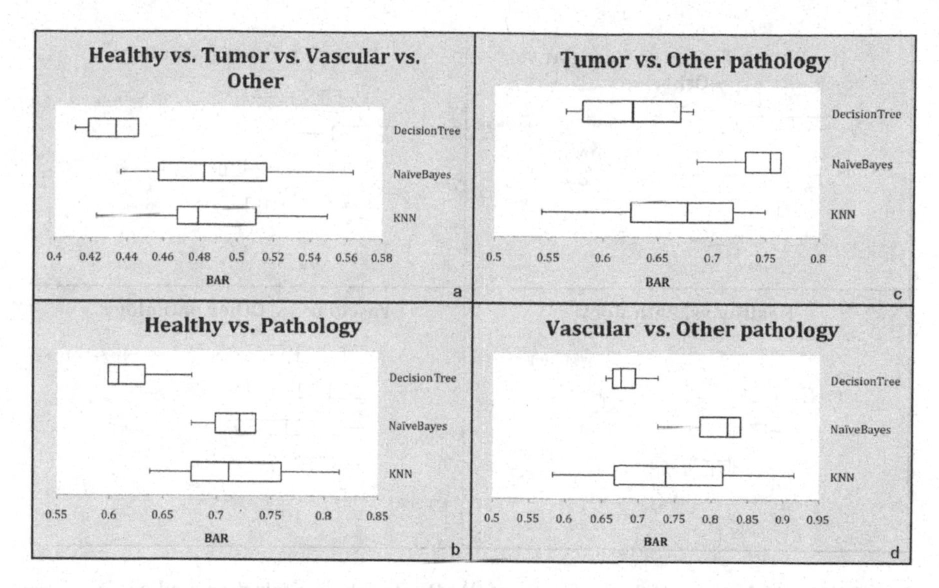

Fig. 2 Box-whisker plots of the performance (BAR) for each classification problem

distance between the unclassified sample and the classified samples. In decision tree, the pruning strategy and information gain ratio as purity measure was used.

Fig. 2 (a) shows the box-whisker plot of the performance for the tumor vs. vascular vs. healthy vs. other pathology problem.

The quality measures present modest values of confidence. This is due to the fact that some classification algorithms deal poorly with problems where the number of possible class values is high. If the spectrum was classified randomly, the classifier had 25% of probability to choose the proper class, since there are four classes. We obtained values around 48%, which is a good value for this problem, but not good enough for medical purposes.

Any N-class classification problem can be transformed into several 2-class problems [21]. Fig. 2 (b) presents the box-whisker plot of the performance for the healthy vs. pathology classification, with 68 cases, of which 17 (25%) are healthy patients and 51 (75%) are patients with pathology. Fig. 2 (c) presents the box-whisker plot of the performance for the tumor vs. other pathology classification, with 51 cases, of which 25 (49%) are patients with tumor and 26 (51%) are patients with other pathology. Fig. 2 (d) presents the box-whisker plot of the performance for the patient with vascular pathology and patient with other pathology classification, with 26 cases of which 14 (54%) are patients with vascular pathology and 12 (46%) are patients with other pathology.

KNN presented the best results for the healthy patient vs. patient with pathology classification. NaïveBayes presented the best results for the other two classifications. These results are much better than the classifier that distinguished the four classes. This happens because the difference between the values of the different metabolites is higher when they are distinguished between two classes.

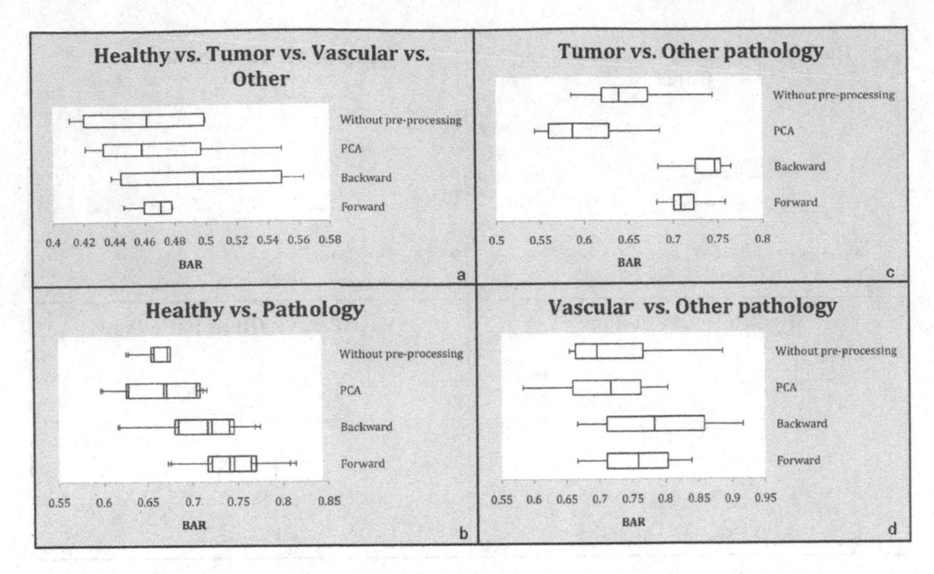

Fig. 3 Box-whisker plot of the performance (BAR) of each data selection method

Almost all worst results of the 2-classes problem classifiers were better than the best results of the 4-classes problem, which is quite obvious since a 4-classes problem is intrinsically more difficult than a 2-classes problem. So, the classifiers for 2-classes problems were able to resolve the classification problems with BAR values around or above 70%, depending of the classification problem.

As mentioned before, the Feature Selection technique has the objective to reduce the dimension and the PCA to compress data. These techniques were applied before the data mining task. In Feature Selection, both forward and backward directions were chosen. In PCA the first principal components that explained 90% of the original values were chosen. Since the results of performance for the discrimination problems were distinct enough, it is more appropriate to analyze them separately.

Fig. 3 (a) presents the box-whisker plot of the performance for each data selection method. The best results were obtained with Feature Selection in the backward direction. This seems to indicate that it isn't necessary all the attributes to solve this discrimination problem because this procedure removes, at each step, the worst attribute remaining in the set. Fig. 3 (b), Fig. 3 (c) and Fig. 3 (d) show the performance results for the 2-classes problems. In all them, the Feature Selection in both directions: forward and backward were the bests. The BAR average was always above 70%. This seems to show that, when the dimensionality is reduced, part of the irrelevant information to solve the discriminations problems is removed.

Even thought there are several published studies where classification algorithms were used to classify the spectra, such as, LDA [9] [22] [23], LS-SVM [9] [23], Decision Tree, KNN [24] and NaïveBayes [25], there is not yet a consensus on which are the best techniques for MRS data processing to be used for optimum classification [26]. In this study KNN, NaïveBayes and Decision tree classification algoritms were used. In all classification problems, NaïveBayes presented the best

BAR average (around 75%). With the exception of the 4-classes problem, the KNN classification method obtained a higher inter-quartile range. This fact contributed for the reduction of BAR average since the inter-quartile range represents half of the samples. The Decision Tree presented the worst results. The recursive partitioning and the sequential nature of the decision tree can make it dependent of the observed data, and even a small change might alter the structure of the tree [27]. This could have been one of the reasons for its performance.

4 Conclusions

The use of KNN, Naïve Bayes and Decision Tree classifiers in identifying four types of patient's spectra was evaluated. Initially the task to distinguish the four types simultaneously was analyzed and then in pair wise tasks (i.e. healthy vs. with pathology, tumor vs. other knowing that they are patients with pathology and vascular vs. other knowing that they are patients with pathology different from tumor). The best results were obtained by the classifiers in pair wise tasks.

The 2-classes problems presented all BAR values around and above 70%. In contrast, the 4-classes problem never achieved values higher than 57%. This may be due to the fact that some algorithms deal poorly with problems where the number of possible class values is high. All N-class problems had similar behavior in relation to data selection. In all trials, Feature Selection in forward or backward direction is a good option as data selection technique. This seems to indicate that the classifiers might not need all attributes in order to correctly discriminate.

For further experiments, it would be necessary to collect more samples of each of the classified classes to perform further research with a more balanced population of cases. Finally, the obtained results leads us to conclude that it is possible to classify magnetic resonance spectra with data mining techniques for further integration in a Clinical Decision Support System with the goal to help in the diagnosis of new cases.

References

1. Bushberg, J.T., Seibert, J.A., Leidholdt, E.M., Boone, J.M.: The Essential Physics of Medical Imaging, 2nd edn. Lippincott Williams & Wilkins, Philadelphia (2002)
2. Sibtain, N.A., Howe, F.A., Saunders, D.E.: The clinical value of proton magnetic resonance spectroscopy in adult brain tumours. Clinical Radiology 62, 109–119 (2007)
3. Callot, V., et al.: 1H MR spectroscopy of human brain tumours: a practical approach. European Journal of Radiology 67, 268–274 (2008)
4. Tate, A.R., et al.: Automated Classification of Short Echo Time in In Vivo 1H Brain Tumor Spectra: A Multicenter Study. Magnetic Resonance in Medicine 49, 29–36 (2003)
5. Übeyli, E.D.: Comparison of different classification algorithms in clinical decision-making. Expert systems 24(1), 17–31 (2007)
6. Eom, J., Kim, S., Zhang, B.: AptaCDSS-E: A classifier ensemble-based clinical decision support system for cardiovascular disease level prediction. Expert Systems with Applications 34, 2465–2479 (2008)

7. Bezabeh, T., et al.: Statistical classification strategy for proton magnetic resonance spectra of soft tissue sarcoma: an exploratory study with potential clinical utility. Sarcoma 6, 97–103 (2002)
8. Siddall, et al.: Magnetic Resonance Spectroscopy Detects Biochemical Changes in the Brain Associated with Chronic Low Back Pain: A Preliminary Report. Anesthesia and Analgesia 102, 1164-1168 (2006)
9. Luts, et al.: A combined MRI and MRSI based multiclass system for brain tumour recognition using LS-SVMs with class probabilities and feature selection. Artificial Intelligence in Medicine 40(2), 87–102 (2007)
10. García-Gómez, et al.: Multiproject-multicenter evaluation of automatic brain tumor classification by magnetic resonance spectroscopy. Magnetic Resonance Materials in Physics 22(1), 5–18 (2009)
11. Siemens, MR Spectroscopy Operator Manual: Version syngo MR 2002B. Erlangen (2002)
12. Han, J., Kamber, M.: Data Mining: Concepts and Techniques. Morgan Kaufmann, San Francisco (2000)
13. Luts, J., et al.: Effect of Feature Extraction for Brain Tumor Classification Based on Short Echo Time 1H MR Spectra. Magnetic Resonance in Medicine 60, 288–298 (2008)
14. Duda, R.O., Hart, P.E., Stork, D.G.: Pattern Classification. Wiley-Interscience, Hoboken (2000)
15. Hastie, T., Tibshirani, R., Friedman, J.H.: The Elements of Statistical Learning: Data Mining, Inference and Prediction. Springer, Heidelberg (2003)
16. Berry, M.J.A., Linoff, G.S.: Data Mining Techniques: For Marketing, Sales, and Customer Relationship Management. Wiley Computer Publishing, Indianapolis (2004)
17. Kantardzic, M.: Data Mining: Concepts, Models, Methods and Algorithms. Wiley-IEEE Press (2002)
18. An, A.: Classification Methods. In: Wang, J. (ed.) Encyclopedia of Data Warehousing and Mining, pp. 144–149. Idea Group Publishing, Hershey (2006)
19. Kertész-Farkas, A., et al.: Benchmarking protein classification algorithms via supervised cross-validation. Journal of Biochemical and Biophysical Methods 70, 1215–1223 (2008)
20. Kohavi, R.: A Study of Cross-Validation and Bootstrap for Accuracy Estimation and Model Selection. In: IJCAI (International Joint Conferences on Artificial Intelligence), Proceedings of the Fourteenth International Joint Conference on Artificial Intelligence, Québec, Canada, August 20-25, Morgan Kaufmann, San Francisco (1995)
21. Guillet, F., Hamilton, H.J.: Quality Measures in Data Mining. Spinger, Warsaw (2007)
22. Lukas, L., et al.: Brain tumor classification based on long echo proton MRS signals. Artificial Intelligence in Medicine 31, 73–89 (2004)
23. Devos, A., et al.: Classification of brain tumours using short echo time 1H MR spectra. Journal of Magnetic Resonance 170, 164–175 (2004)
24. Manocha, S., Girolami, M.A.: An empirical analysis of the probabilistic K-nearest neighbour classifier. Pattern Recognition Letters 28, 1818–1824 (2007)
25. Kazmierska, J., Malicki, J.: Application of the Naïve Bayesian Classifier to optimize treatment decisions. Radiotherapy and Oncology 86, 211–216 (2008)
26. Opstad, et al.: Linear discriminant analysis of brain tumour 1H MR spectra: a comparison of classification using whole spectra versus metabolite quantification. NMR in Biomedicine 20, 763–770 (2007)
27. Giudici, P.: Applied Data Mining: Statistical Methods for Business and Industry. Wiley, England (2003)

A Bio-inspired Ensemble Model for Food Industry Applications

Bruno Baruque, Emilio Corchado, and Jordi Rovira

Abstract. This paper presents a soft computing robust solution for the food industry field with the aim of analysing the olfactory properties of Spanish dry-cured ham. A novel topology preserving version of the Visualization Induced SOM (ViSOM), based on the application of the Weighted Voting Superposition (WeVoS) summarization algorithm, is presented in order to calculate the best possible visualization of the internal structure of a datasets. The results obtained by this novel model are compared with the ones obtained by its single version -ViSOM- and versus the well-known SOM and WeVOS-SOM. The results clearly demonstrate how the WeVoS-ViSOM outperforms the rest of models.

1 Introduction

Soft computing has been successfully applied in many areas, such as biology, finance, and marketing [7]. In this study, soft computing techniques are apply to the field of food industry, where, ensuring the taste and quality of end products is a critical issue. The quality of these products depends in a complex way on many factors. In order to effectively control food taste and conditions, this research aims at implementing a system that classifies the quality and taste of Spanish ham samples. In this system we use soft computing techniques and exploit large amounts of data collected throughout an "Electronic Nose" (or "e-nose"). This paper investigates

Bruno Baruque
University of Burgos
e-mail: bbaruque@ubu.es

Emilio Corchado
University of Salamanca
e-mail: escorchado@usal.es

Jordi Rovira
University of Burgos
e-mail: jrovira@ubu.es

E. Corchado et al. (Eds.): SOCO 2010, AISC 73, pp. 209–217.
springerlink.com

requirements on soft computing techniques as topology preserving models and fusion theory. Results of a preliminary case study show that soft computing is a promising approach as part of early systems in food industry.

Topology preserving maps [3, 4] are often used for data visualization and inspection tasks. This interesting feature can assist human operators in classification tasks, such as the one presented in this study relating to the olfactory properties of Spanish dry-cured ham. Other features are pattern recognition and automated classification, inherent to many of the unsupervised learning techniques, which are especially relevant in the present application. These models are given enhanced stability in this study through the use of a novel ensemble summarization algorithm, called Weighted Voting Superposition (WeVoS). A combination of an electronic device for the analysis of volatile compounds and a novel ensemble summarization algorithm for topology preserving mapping algorithms is used to study a wide variety of samples of Serrano Hams, in order to test whether this procedure is able to discriminate, in an easy and reliable way, between hams with different olfactory characteristics.

The rest of this paper is organized as follows: Section 2 and Section 3 present the soft computing models used in this research. Section 4 outlines the obtaining and pre-processing of the data of the industrial case study presented, while Section 5 describes the experiments and results. Finally, Section 6 presents the conclusions and future lines of research.

2 Topology Preserving Maps and Quality Measures

2.1 The Self-organizing Map

Topology preserving mapping comprises a family of techniques with a common target: to produce a low-dimensional representation of the training samples that preserves the topological properties of the input space. From among the various techniques, the best known is the Self-Organizing Map (SOM) algorithm [3, 4]. SOM aims to provide a low-dimensional representation of multi-dimensional data sets while preserving the topological properties of the input space. The SOM algorithm is based on competitive unsupervised learning; an adaptive process in which the neurons in a neural network gradually become sensitive to different input categories, which are sets of samples in a specific domain of the input space [5]. The update of neighbourhood neurons in SOM is expressed as:

$$w_k(t+1) = w_k(t) + \alpha(t)\eta(v,k,t)(x(t) - w_k(t)) \tag{1}$$

where, x denotes the network input, w_k the characteristics vector of each neuron; α, is the learning rate of the algorithm; and $\eta(v,k,t)$ is the neighbourhood function, in which v represents the position of the winning neuron (BMU) in the lattice, and k the positions of the neurons in its neighbourhood.

2.2 The Visualization Induced SOM

An interesting extension of this algorithm is the Visualization Induced SOM [9, 8] proposed to directly preserve the local distance information on the map, along with the topology. The ViSOM constrains the lateral contraction forces between neurons and hence regularizes the inter-neuron distances so that distances between neurons in the data space are in proportion to those in the input space, as it is expressed in Eq. :

$$w_k(t+1) = w_k(t) + \alpha(t)\eta(v,k,t)\left[(x(t) - w_v(t)) + (w_v(t) - w_k(t))\frac{d_{vk} - \Delta_{vk}\lambda}{\Delta_{vk}\lambda}\right] \tag{2}$$

where, d_{vk} and Δ_{vk} are the distances between neurons in the data space v and k on the unit grid or map, respectively, and λ is a positive pre-specified resolution parameter. It represents the desired inter-neuron distance -of two neighbouring nodes- reflected in the input space.

The difference between the SOM and the ViSOM hence lies in the update of the weights of the neighbours of the winner neuron as can be seen from Eq. 1 and Eq. 2.

2.3 Measures for Assessing the Quality of a Map

This subsection presents some of the measures that have been proposed in literature to compare the maps resulting of two different trainings. As any unsupervised learning technique, to obtain a definitive overall measure of the performance of the model is a very complicated problem. In the case of the present study, two of the most simple of them have been used to analyze the outcome of the ensemble summarization algorithm once applied to the ensemble of topology preserving models:

Classification Error: based on classifying each new sample as part of the class that most consistently recognized the BMU for that sample [6]. In the case of the use of the model for the practical purpose of the food industry suggested in this work, the classification of samples is an interesting added feature.

Mean Square Quantization Error: computed by determining the average distance of the test data set entries to the cluster centroids by which they are represented.

3 Ensemble Summary of Visualization Models

The idea behind the novel fusion algorithm, WeVoS, is to obtain a final map keeping one of the most important features of these types of algorithms: its topological ordering. WeVoS, which is presented in [1]; has been previously used in other practical applications [2]; but in this study is applied for the first time to the ViSOM and in the field of the food industry.

It is based on the calculation of the "quality of adaptation" of a homologous unit of different maps, in order to obtain the best characteristics of the vector in each of the units that make up the final map. This calculation is performed as follows:

$$V_{p,m} = \frac{\sum b_{p,m}}{\sum_{i=1}^{M} b_{p,i}} \cdot \frac{\sum q_{p,m}}{\sum_{i=1}^{M} q_{p,i}} \quad (3)$$

Briefly, the WeVoS algorithm functions in the following way: first of all an ensemble of maps is trained. Then, the chosen quality/error measure is calculated for each of the neurons in all the ensemble maps. The fused map is initialized by calculating the centroids of the neurons in the same position of all the maps, that is, by calculating the superposition of the ensemble. For each of the neurons in the fused map, the average neuron quality as well as the number of total samples recognized in that position for the ensemble maps are calculated. The weight of the vote for each neuron can be calculated with this information by using Eq. 3. To modify the position of the neuron in the fused map, the weights of each of the neurons in that position are fed to the final map. The learning rate in each case will be the weight of the vote for that neuron. A complete description of the WeVoS algorithm can be found in Algorithm 1.

Algorithm 1. Weighted Voting Summarization algorithm

Input: Set of trained topology-preserving maps: $M_1 ... M_n$, training data set: S
Output: A final fused map: M_{fus}

1: Select a training set $S = \langle (x_1, y_1) \dots (x_m, y_m) \rangle$
2: train several networks by using the bagging meta-algorithm : $M_1 ... M_n$
3: **procedure** WEVOS($M_1 ... M_n$)
4:    **for all** map $M_i \in M_n$ **do**
5:       calculate the quality/error measure chosen for ALL neurons in the map
6:    **end for**
▷ These two values are used in Eq. 3
7:    calculate an accumulated total of the quality/error for each position $Q(p)$
8:    calculate recognition rate for each position $B(p)$.
9:    **for all** unit position p in M_i **do**
10:       initialize the fused map (M_{fus}) by calculating the centroid (w_c) of the neurons of all maps in that position (p)
11:    **end for**
12:    **for all** map $M_i \in M_n$ **do**
13:       **for all** unit position p in M_i **do**
14:          calculate the vote weight (V_{p,M_i}) using Eq. 3.
15:          feed the weights vector of neuron w_p into the fused map (M_{fus}) as if it was an input to the network.
The weight of the vote (V_{p,M_i}) is used as the learning rate (α).
The position of that neuron (p) is considered as the position of the BMU (v). ▷ This causes the neuron of the fused map (w_p^*) to approximate the neuron of the composing ensemble ($w_{p,m}$) according to the quality of its adaptation.
16:       **end for**
17:    **end for**
18: **end procedure**

---

4 A Food Industry Application Case Study

Several Spanish hams of different qualities and origins were used in this case study. The data sets consisted of measurements taken from seven types of Spanish dry-cured ham from among the various brands available on the Spanish market. The samples also included some that were tainted and/or that had a rancid/acidic taste. The tainted samples were randomly taken from among all the different quality types and origins of hams. The commercial brands of the hams in the samples were not taken into account in this study. In this case the e-nose was used to measure the odour of the ham samples. The data collected was presented to the ensemble summarization algorithm of topology preserving maps, WeVoS-ViSOM, in order to achieve a simple and reliable device for testing and analyzing the olfactory properties of the hams.

The odour recognition process followed in this research may be summarized as follows:

1. The sample is heated for a given time to generate volatile compounds in the head space of the vial containing the sample.
2. The gas phase is transferred to a detection device which reacts to the presence of molecules.
3. The differences in sensor reactions are recorded using statistical calculation techniques to classify the odours.

The readings taken by each sensor are separated and stored in a simple database for further study. In this study the analyzes are performed using an E-Nose α-FOX 4000 (Alpha M.O.S., Toulouse, France) with a sensor array of 18 metal oxide sensors.

After having obtained the readings for each sample of cut ham taken from the 18-sensor array in the electronic nose, they are stored in a database along with the corresponding results of the sensory evaluation by the professional testers. These results are normally more detailed, but were restricted in this initial study to three possible values: “unspoilt”, “rancid/acid” and “tainted”. Thus, the final data set consisted of readings taken from a total of 154 samples of ham, the readings on each ham being composed of 18 different variables measured over three possible categories.

5 Experimental Results

In order to analyze the validity of the idea of using the topology preserving maps as a clearer way to present the results of the chemical analysis of the e-nose to a human user, two types of experiments are presented in this research. In both experiments we compare the single algorithms (SOM and ViSOM) with their ensemble summary counterpart, as the final idea is to study the benefits of the use of an ensemble of those algorithms.

After having obtained the readings for each sample of cut ham taken from the 18-sensor array in the electronic nose, they are stored in a database along with the corresponding results of the sensory evaluation by the professional testers. These results are normally more detailed, but were restricted in this initial study to three possible values: “unspoilt”, “rancid/acid” and “tainted”. Thus, our final data set consisted of readings taken from a total of 154 samples of ham, the readings on each ham being composed of 18 different variables measured over three possible categories.

Figure 1 shows the ham data set as the map obtained by each of the topology preserving models discussed: SOM, ViSOM and their ensemble counterparts: WeVoS-SOM and WeVoS-ViSOM. Triangles represent unspoilt samples, squares rancid/acid samples and circles tainted samples.

It is worth noting some interesting features in those maps. First of all, the WeVoS algorithms help to obtain more compact maps. Comparing Fig. 1a and Fig. 1c and also Fig. 1b and Fig. 1d it is easy to see that WeVoS models obtain maps with less space (i.e. “dead neurons”) in the center of the figures, obtaining more compact and clear maps, with less wasted blank space. Specially ViSOM and WeVoS-ViSOM

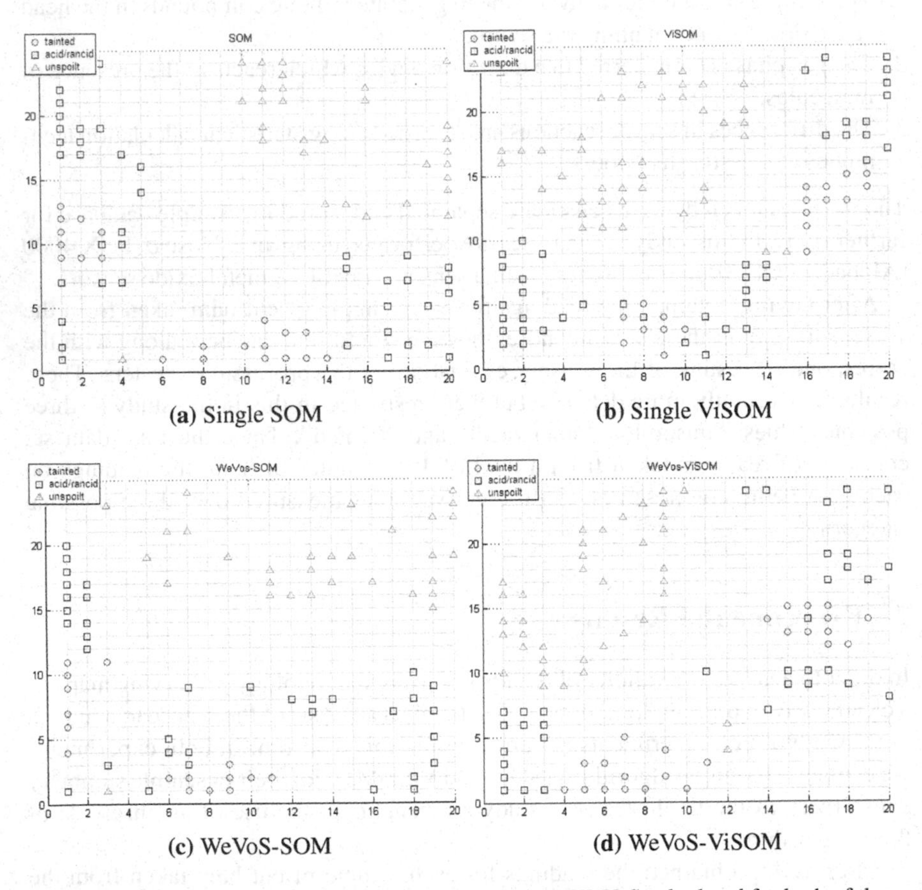

(a) Single SOM

(b) Single ViSOM

(c) WeVoS-SOM

(d) WeVoS-ViSOM

Fig. 1 Maps generated form both single models and the WeVoS calculated for both of them obtain clearer images than the SOM and WeVoS-SOM. For example, ViSOM models (Fig. 1b and Fig. 1d) are capable of identify a small group of "unspoilt" samples -that are separated from the main group due to their special characteristics that can be taken for "acid/rancid" samples- using 2 units, which is an improvement over the Single SOM (Fig. 1a) -no units- and the WeVoS-SOM (Fig. 1c) -only one unit-. Even between those two models, the WeVoS-ViSOM obtains a clearer map, considering that groups are more compact and clearly separated. Seeing the bottom of the images, in the WeVoS-ViSOM (Fig. 1d) obtains much less messed groups for the "tainted" and "acid/rancid" samples.

The second experiment consisted in using a fixed number for the size of the ensemble of maps -5 maps- determined experimentally, while decreasing the number of samples used to train them. The initial step experiment was performed with the complete data set and in each progressive step the 1/5 of the original size of samples were randomly chosen to be removed from the data set. This is done in order to analyze which is the effect of the addition of instability to the data set by making it more difficult to adapt to, as it contains less samples to train.

Being the topology preserving maps quite error tolerant algorithms, the effect of reducing the number of samples does not have a clear effect in the calculated measures. This means that the data set does not include any outliers or noisy samples, with very different characteristics from the rest. As there are no outliers present in the data set, the decrease in the number of samples used does not really alter its internal structure, so tests tend to be quite stable in every size of data set.

The classification error of the WeVoS models (Fig. 2a) is clearly lower compared with the single algorithms regardless the size of the data set. For the Quantization Error (Fig. 2b), the fact of having a lower size for the data set affects the error that can be measured. In this case, although ViSOM and WeVoS-ViSOM obtain lower error, they are not as clearly superior as in previous experiment. It is interesting to see how the use of the WeVoS algorithm make the unstable algorithm of the

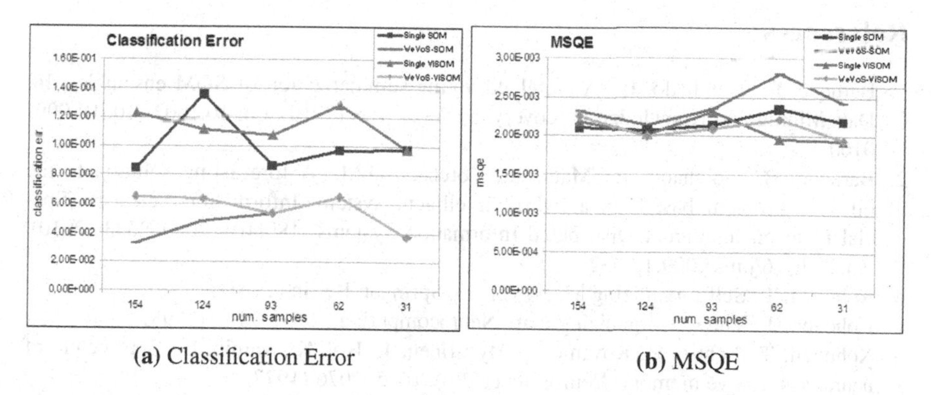

(a) Classification Error (b) MSQE

Fig. 2 Map quality measures when varying the number of data samples used to train the ensemble

ViSOM behave in a less unstable way. The measures, that exhibit a certain degree of instability for single models, behave in a less variable way when measured for the summarized ensembles using WeVoS.

6 Conclusions

In this paper a soft computing technique for presenting the results of a complex chemical analysis, from a food industry case study; in a simple and understandable way to a human expert has been presented and analyzed. The analysis steps include the chemical analysis performed in a device designed for the pourpose and the posterior analysis of the data collected by means of topology preserving maps. A summarization algorithm for this type of topology preserving algorithms is presented and used to add an extra fine tuning to the final results and the stability which they mostly lack of. An analysis and comparison of the use of this algorithm with two topologies preserving models is included, being this the first time that the algorithm has been used in conjunction with the Visualization Induced SOM (ViSOM).

Some of the future work includes the application of the WeVoS algorithm to other topology preserving models to test its performance in conjunction with the e-nose. A more complete array of quality measures will be used in order to study in a more detailed way the effect of the use of ensembles with these topology preserving models. Also, the use of some other ensemble algorithms such as boosting or leveraging for the training of the ensembles will be explored.

Acknowledgements. This research is funded through the Junta of Castilla and León (BU006A08) and the Spanish Ministry of Education and Innovation (CIT-020000-2008-2 and CIT-020000-2009-12). The authors would also like to thank the vehicle interior manufacturer, Grupo Antolín Ingenieria S.A. for supporting the project through the MAGNO2008 - 1028 - CENIT Project funded by the Spanish Ministry of Science.

References

1. Baruque, B., Corchado, E.: A weighted voting summarization of SOM ensembles. In: Data Mining and Knowledge Discovery, Springer, U.S (2010), doi:10.1007/s10618-009-0160-3
2. Baruque, B., Corchado, E., Mata, A., Corchado, J.M.: A forecasting solution to the oil spill problem based on a hybrid intelligent system. Information Sciences (Special Issue on Intelligent Distributed Information Systems) 180(10), 2029–2043 (2010), doi:10.1016/j.ins.2009.12.032
3. Kohonen, T.: Self-Organizing Maps, vol. 30. Springer, Berlin (1995)
4. Kohonen, T.: The self-organizing map. Neurocomputing 21(1-3), 1–6 (1998)
5. Kohonen, T., Lehtio, P., Rovamo, J., Hyvarinen, J., Bry, K., Vainio, L.: A principle of neural associative memory. Neuroscience 2(6), 1065–1076 (1977)

6. Kraaijveld, M.A., Mao, J., Jain, A.K.: A nonlinear projection method based on kohonen's topology preserving maps. IEEE Transactions on Neural Networks 6(3), 548–559 (1995)
7. Van der Vorst, J.G.: Performance measurement in agrifood supply chain networks: an overview (2005)
8. Yin, H.: Data visualisation and manifold mapping using the ViSOM. Neural Networks 15(8-9), 1005–1016 (2002)
9. Yin, H.: ViSOM - a novel method for multivariate data projection and structure visualization. IEEE Transactions on Neural Networks 13(1), 237–243 (2002)

Implementation of a New Hybrid Methodology for Fault Signal Classification Using Short -Time Fourier Transform and Support Vector Machines

Tribeni Prasad Banerjee, Swagatam Das,
Joydeb Roychoudhury, and Ajith Abraham

Abstract. Increasing the safety of a high-speed motor used in aerospace application is a critical issue. So an intelligent fault aware control methodology is highly research motivated area, which can effectively identify the early fault of a motor from its signal characteristics. The signal classification and the control strategy with a hybrid technique are proposed in this paper. This classifier can classify the original signal and the fault signal. The performance of the system is validated by applying the system to induction motor faults diagnosis. According to our experiments in BLDC motor controller results, the system has potential to serve as an intelligent fault diagnosis system in other hard real time system application. To make the system more robust we make the controller more adaptive that give the system response more reliable.

Keywords: Real time system, Support Vector Machine, Sort Time Fourier Transform, embedded system, Brash less Direct Current Motor, Intelligent interactive control, supervisory control, signal classification, Fault Classifier, Mechatronics, fault diagnosis.

Tribeni Prasad Banerjee · Joydeb Roychoudhury
Embedded System Laboratory, Central Mechanical Engineering Research Institute, Durgapur-713209, India
e-mail: t_p_banerjee@yahoo.com, jrc@cmeri.res.in

Swagatam Das
Electronics and Telecommunication Engineering Department Jadavpur University, Jadavpur-700035, India
e-mail: swagatamdas19@yahoo.co.in

Ajith Abraham
Machine Intelligence Research Labs (MIR Labs), Scientific Network for Innovation and Research Excellence, USA
e-mail: ajith.abraham@ieee.org

E. Corchado et al. (Eds.): SOCO 2010, AISC 73, pp. 219–225.
springerlink.com

1 Introduction

Fault diagnosis is very important for safe operation and preventing rescue [1]. It is preferable to find fault before complete system failure. A system with incipient fault but it will lead to a system to a catastrophic failure causing downtime and large losses. Fault diagnostics is a kind of signal classification as far as its essence is concerned. Support vector machines (SVM) have a good generalization capability even in the small-sample cases of classification and have been successfully applied in fault detection and diagnosis [2][3]. SVMs [4-7] provide efficient and powerful classification algorithms that are capable of dealing with high-dimensional input features and with theoretical bounds on the generalization error and sparseness of the solution provided by statistical learning theory [4]. Classifiers based on SVMs have few free parameters requiring tuning, are simple to implement, and are trained through optimization of a convex quadratic cost function, which ensures the uniqueness of the SVM solution. Furthermore, SVM-based solutions are sparse in the training data and are defined only by the most "informative" training points.

In control and automata theory, a state is defined as a condition. That characterizes a prescribed relationship of input, output and input to the next state. Thus knowledge of the state of a system gives better understanding through observation and we can subsequently take necessary steps for controlling the system, for example, stabilizing a system using state feedback. This implies that the current state of a system determines the future plan of action. So it becomes extremely important to devise some methodology by which the state of a system can be observed correctly within a stipulated time frame. For example a particular state of a feedback control system cannot be observed directly due to its own compensator which is one of the limitations of the feedback controller. A separate hardware with very low observation latency period [6] is necessary to observe this unobserved state which is called a sub state or internal state directly related to the health of the system. In order to make system fault aware a sub state observer is conceived whose execution time of the event sequences (necessary for sub state observation) is much faster and due to this high speed execution and subsequent scheduling technique, it can observe the sub state behavior of a system and thus differentiate between two almost similar states with different end effects.

In this paper, we proposed an incipient signal classification which is optimally separated by SVM and after classification of the signal or a optimal threshold value a controller strategy is implanted into a evolvable hardware which gives a hard and real time responsive or responsive system.

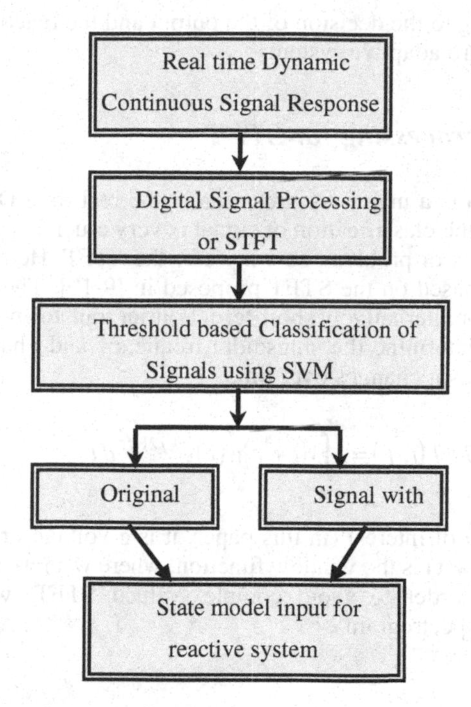

Fig. 1 A generic Scheme for signal classification and reactive system and their consequences

The basic scheme of Signal classification is illustrated in figure 1.

2 Principal of Operation Signal Classification

2.1 Real Time Dynamic Continuous Signal Response

Hard real time system is that system which has a critical time limit [10] or a reactive system where the system inputs are dynamically changes with time. Within that time duration the system has to response otherwise the system is going to a catastrophic loss or high probability for human death. The real time systems are basically two types one is reactive system and another is embedded system. A reactive system is always react with the environment (online aircraft valve signal monitoring from a actuator signal) and another is embedded system which is used to control specialized hardware that is installed within a larger system (such as a microprocessor that controls anti-lock brakes in an automobile) In our system is more reactive with the environments. Here in our proposed system the online dynamic continuous signals is comes and we capture the signal and transform it and pass through into our classifier and take the decision. The EFSM model [6] does the real time operation within a few micro seconds. This makes the system first responsive according to the decision of the output and the reactive system makes a more intelligent neuro adaptive system.

2.2 Signal Preprocessing for STFT

Signal classification is a major research area in the real time DSP, because of the real time operation the classification of signal is very crucial issue. Even the signal classification has lots of problem, so we prefer the STFT. Here we briefly review the chaos detector based on the STFT proposed in [9-10]. The short-time Fourier transform (STFT), or alternatively short-term Fourier transform, is a Fourier-related transform used to determine the sinusoidal frequency and phase content of local sections of a signal as it changes over time.

$$STFT(t,f) = \int_{-\alpha}^{\alpha} x(t+\tau)w(\tau)e^{-j2\Pi f\tau}d\tau \quad (1)$$

Where x(t) is signal of interest (in this paper, it is a voltage or a current from the BLDC motor), and w(τ) is the window function, where w(τ)=0 for |τ|>T/2 and T is window width. In order to avoid complex-valued STFT, we use its squared magnitude ,i.e.,the spectrogram $SPEC(t,f) = |STFT(t,f)|^2$. As shown in figure 2

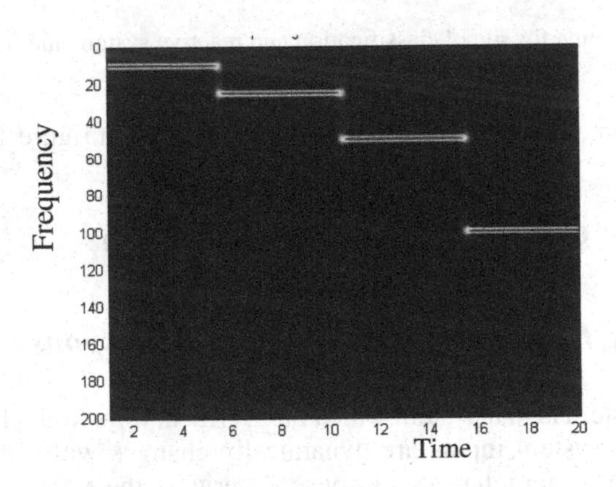

Fig. 2 Original Signal after the STFT transformed

2.3 Signal Classification of Using SVM

Support vector machines (SVM) [10] have been widely used in the fault classification and diagnosis of machines in the past few years, especially in the fault diagnosis [11], for dynamic procedures such as the starting, the stopping, and the changing of working mode. SVM certainly outperform some other artificial intelligence methods because it is able to maximize the generalization performance of a trained classifier. This may not be easily achieved by using HMM or ANN, because ANN are prone to having a higher specificity and a lower sensitivity and HMM are prone to having a higher sensitivity and a lower specificity. SVM aims at the optimal solution in the available information rather than just the optimal solution when the sample number tends towards infinitely large. It has a good generalization when the samples are few, so it is especially fit for classification, forecasting and estimation in small-sample cases such as fault diagnosis, in most cases whose bottleneck problem is the lack of fault samples.

Once the frequency is separated by STFT technique, then the signals are taken into the classification stage, here according to the maximized threshold value are calculated by SVM. Because that optimum threshold value is given to the state machine for taken into the real time decision processes. Depending upon the magnitude value the threshold values are set to a specific signal and the different signal magnitude are set to a different class. A signal which is exceed a specific threshold value it is classified as high risk and other which are not exceed is fallen into other class. This threshold value is set dynamically and depending upon the timed model components is already classified form the inputs.

SVM [10], as well as neural networks, have been extensively employed to solve classification problems due to the excellent generalization performance on a wide range problem. In machine fault diagnosis, some researchers have employed SVM as a tool for classification of faults. For example, ball bearing faults [9], gear faults [10], condition classification of small reciprocating compressor, and induction motor [11]. In this paper, we also proposed a incipient faults identification in dynamic signal diagnosis system of induction motor based on start-up transient current signal using component analysis and SVM.

In SVM, the original input space is mapped into a high-dimensional dot product space called a feature space, and in the feature space. SVM' s have the potential to handle very large feature spaces, because the training of SVM is carried out so that the dimension of classified vectors does not have as a distinct influence on the performance of SVM as it has on the performance of conventional classifiers. That is why it is noticed to be especially efficient in large classification problems [12].This will also benefit in fault classification, because the number of features to be the basis of fault diagnosis may not have to be limited. Also, SVM-based classifiers are claimed to have good generalization properties compared to conventional classifiers, because in training the SVM classifier, the so-called structural misclassification risk is to be minimized, whereas traditional classifiers are usually trained so that the empirical risk is minimized. When a signal falls outside the clusters, it is tagged as a potential motor failure [13]. Since a fault condition is not a spurious event it can degrade the system, the postprocessor alarms the user only after multiple indications of a potential failure have occurred. In this way, the time duration modeling of the State machine is incorporated into the monitoring system and protects the reactive system by alarming on random signals that have been identified as in our system from the mixed signal the signal of interest has been separated from the injected fault signal, shown in the figure 3 (a) & (b).

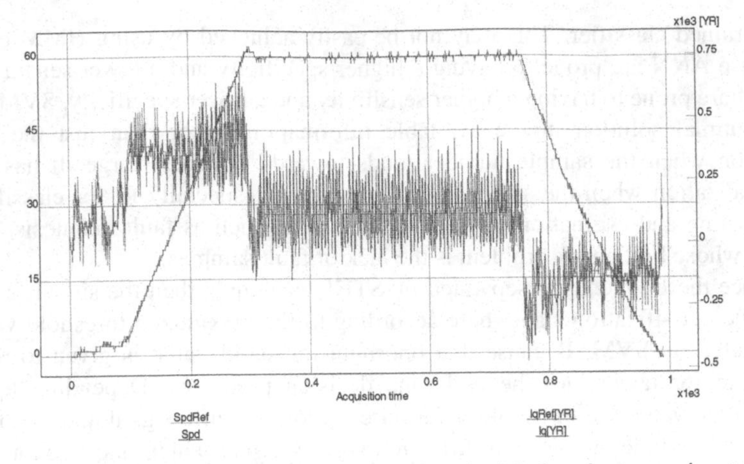

Fig 3.Original Signal and the fault signal are both before the system input

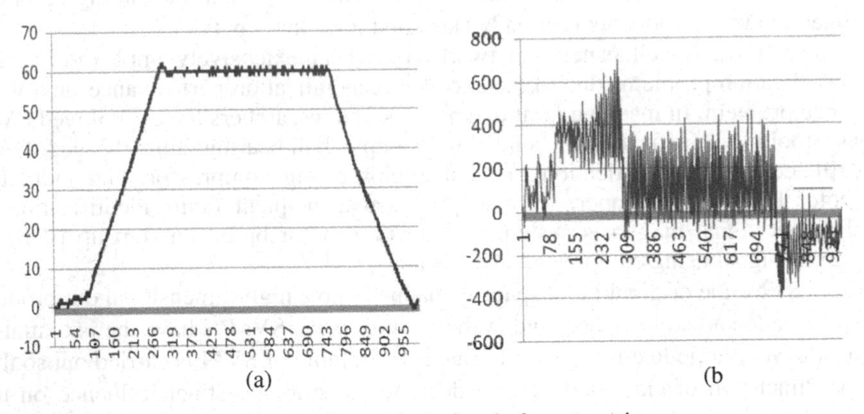

Fig. 3 The actual signal (a, b) and injected fault signal after separation

3 Experimental Set Up and System Description

We used the TMS120F2812, which is a 12-bit Fixed-point DSP of Harvard Bus Architecture. It has separated Data Bus and Program Bus. It also supports atomic operations, 12 x 12 bit MAC operations and fast interrupts response. It is able to control the motor easily with TMS120F2812 because it has two event managers, which include capture function, PWM function, and QEP (Quadrature -Encoder Pulse) function. Since it also supports variable serial port peripherals such as CAN, SCI, and SPI[16], it is possible to communicate with external device and exchange control signal and data. The prototype is designed with a control circuit based on the "MSK2812 Kit C Pro VS" (from Technosoft S.A.). Operating at a 190-MHz frequency, with double event-manager signals, increased internal memory, etc., this new kit offers all features needed for an advanced digital control implementation. All communication between the PC and the DSP board is done through the RS-192 interface using a real-time serial communication monitor with download/upload functions, debug and inspect facilities [14, 15]. PROCEV28x, graphical evaluation/analysis software for the specific peripherals is embedded in the TMS120F2812 DSP.

4 Conclusions

The experiments are done in the laboratory. In our proposed system the intelligent signal classifier classifies the original signal from the fault signal. At the first level, the classifier separates the signal by STFT methods, and the time duration based model classified the signal very smoothly as depicted in Figure 3. We believe that the system has potential to serve an intelligent fault diagnosis system in other hard real time system applications also.

References

[1] Yongming, Y., Bin, W.: A review on induction motor online fault diagnosis. In: The 3rd International Power Electronic And Motion Control Conference 2000 (IPEMC), pp. 1353–1358. IEEE, Los Alamitos (2000)

[2] Matthias, P., Stefan, O., Manfred, G.: Support Vector Approaches for Engine Knock Detection. In: International Joint Conference on Neural Networks, pp. 969–974. IEEE Press, Washington (1999)

[3] Chiang, L.H., Kotanchek, M.E., Kordon, A.K.: Fault diagnosis based on fisher discriminate analysis and support vector machines. Computers and Chemical Engineering 28, 1389–1401 (2004)

[4] Scholkopf, B., Smola, A.: Learning With Kernels. MIT Press, Cambridge (2002)

[5] Herbrich, R.: Learning Kernel Classifiers: Theory and Algorithms. MIT Press, Cambridge (2002)

[6] Mall, R.: Real time systems. In: Theory and practice, Pearson Publication, London (2007)

[7] Alur, R., Dill, D.: The Theory of Timed Automata. Theoretical Computer Science 120, 143–235 (1994)

[8] Jack, L.B., Nandi, A.K.: Fault detection using support vector machines and artificial neural network, augmented by genetic algorithms. Mechanical System Signal Process 14, 373–390 (2002)

[9] Boashash, B. (ed.): Time frequency Signal Analysis and Applications. Elsevier, Amsterdam (2003)

[10] Vapnik, V.N.: The nature of statistical learning theory. Springer, New York (1999)

[11] Yang, B.S., et al.: Fault diagnosis of rotating machinery based on multi-class support vector machines. Journal Mechanical Science Technology 19, 845–858 (2005)

[12] Yang, B.S., Hwang, W.W., Kim, D.J., Tan, A.C.: Condition classification of small reciprocating compressor for refrigerators using artificial neural networks and support vector machines. Mechanical System Signal Process 19, 371–390 (2005)

[13] Ma, X.X., Huang, X.Y., Chai, Y.: PTMC classification algorithm based on support vector machines and its application to fault diagnosis. Control and Decision 14, 212–284 (2003) (in Chinese)

[14] Texas Instruments, TMS120F2812 Digital Signal Processors Data Manual (2005)

[15] Texas Instruments, TMS120F28x DSP Enhanced Controller Area Network (e CAN) Reference Guide (2005)

Advances in Clustering Search

Tarcisio Souza Costa, Alexandre César Muniz de Oliveira,
and Luiz Antonio Nogueira Lorena

Abstract. The Clustering Search (*CS) has been proposed as a generic way of combining search metaheuristics with clustering to detect promising search areas before applying local search procedures. The clustering process may keep representative solutions associated to different search subspaces. Although, recent applications have reached success in combinatorial optimisation problems, nothing new has arisen concerning diversification issues when population metaheuristics, as evolutionary algorithms, are being employed. In this work, recent advances in the *CS are commented and new features are proposed, including, the possibility of keeping population diversified for more generations.

1 Introduction

The Clustering Search (*CS) has been proposed as a generic way of combining search metaheuristics with clustering to detect promising search areas before applying local search procedures [1, 2]. This generalized approach is achieved both by the possibility of employing any metaheuristic and by also applying to combinatorial and continuous optimisation problems. Clusters of mutually close solutions hopefully can correspond to relevant areas of attraction in the most of search metaheuristics, including EAs.

Tarcisio Souza Costa
Universidade Federal do Maranhão, São Luís MA Brasil
e-mail: priestcp@gmail.com

Alexandre César Muniz de Oliveira
Universidade Federal do Maranhão, São Luís MA Brasil
e-mail: acmo@deinf.ufma.br

Luiz Antonio Nogueira Lorena
Instituto Nacional de Pesquisas Espaciais
e-mail: lorena@lac.inpe.br

E. Corchado et al. (Eds.): SOCO 2010, AISC 73, pp. 227–235.
springerlink.com

Relevant search areas can be treated with special interest by the algorithm as soon as they are discovered. The clusters work as sliding windows, framing the search areas and giving a reference point (center) to problem-specific local search procedures. Furthermore, the cluster center itself is always updated by incoming inner solutions. The center update is called assimilation [1, 3].

This basic idea was employed to propose the Evolutionary Clustering Search (ECS) early applied to unconstrained continuous optimisation [1]. Posteriorly, the search guided by clustering was extended to a GRASP (Greedy Randomized Adaptive Search Procedure [4]) with VNS (Variable Neighborhood Search [5]), and applied to Prize Collecting Traveling Salesman Problem (PCTSP) [3]. Despite the expected and desirable population convergence, it must occur in a controlled way, avoiding diversification loss and, in consequence, a poor overall performance. *CS has an inherent potential of controlling the convergence not explored yet: search areas may be monitored not allowing oversearch in a promising area.

This paper is devoted to discuss the recent *CS applications in optimisation problems, focusing how its components were implemented for each application. Other issue treated in this work is concerning population diversification. New features to improve *CS are proposed at last. The remainder of this paper is organized as follows. Section 2 describes the concepts behind *CS's components. Recent advances in each component are described in sections 2.1, 2.2, 2.3 and 2.4, including the now proposed diversification mechanism adequate for the approach improvement. The findings and conclusions are summarized in section 3.

2 Clustering Search Foundations

The *CS attempts to locate promising search areas by framing them by clusters. A cluster can be defined as a tuple $\mathscr{G} = \{c, r, s\}$, where c and r are the *center* and the *radius* of the area, respectively. The radius of a search area is the distance from its center to the edge. There could exist different *search strategies* s associated to the clusters. Initially, the center c is obtained randomly and progressively it tends to slip along really promising sampled points in the close subspace. The total cluster volume is defined by the radius r and can be calculated, considering the problem nature. It is important that r must define a search subspace suitable to be exploited by the search strategies s associated to the cluster.

For example, in unconstrained continuous optimisation, it is possible to define r in a way that all Euclidean search space is covered depending on the maximum number of clusters [1]. In combinatorial optimisation, r can be defined as the number of movements needed to change a solution into another. The neighborhood is function of some distance metric related to the search strategy s [1]. *CS can be splitted off in four independent parts: a search metaheuristic (SM); an iterative clustering (IC) component; an analyzer module (AM); and a local searcher (LS).

2.1 Search Metaheuristic

SM component works as a full-time solution generator, according to its specific search strategy, *a priori*, performing independently of the remaining parts, and manipulating a set of $|P|$ solutions. For example, early implementations of Evolutionary Clustering Search (ECS) employ a real-coded steady-state genetic algorithm as SM component in which a sampling mechanism in selection [1] or updating [1] is used to fed the clustering process. Non-population approaches, as Variable Neighborhood Search (VNS) [6], Simulated Annealing (SA) [7] and Greedy Randomized Adaptive Search Procedure [8] have been employed as SM component to solve combinatorial optimisation as well.

The VNS with Clustering Search (VNS-CS) always executes VND local search, differently of the Greedy Randomized Adaptive Clustering Search GRACS, so named in allusion to a special form to gather *CS with GRASP [8]. In the GRACS the full-time solution generator is a greedy constructive phase, controlled by the parameter α that indicates how greedy a given solution must be built. The improvement phase of a typical GRASP may be computational costly. A supposed advantage of the GRACS is rationally apply costly local search algorithms only to promising areas, represented by cluster centers [8]. Figure 1 illustrates the structural difference between GRASP and GRACS. One can observe that the intrinsic improvement mechanism of GRASP is called according the *CS's criteria.

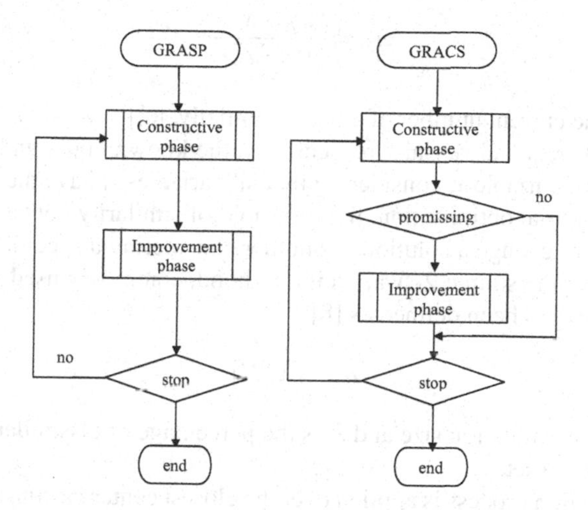

Fig. 1 Difference between GRASP and GRACS

2.2 Iterative Clustering

IC component aims to gather similar solutions into groups, maintaining a representative cluster center for them. To avoid extra computational effort, IC is fed, progressively, by solutions generated in each regular iteration of SM. A maximum number of clusters $\mathcal{NC}$ is an upper bound value that prevents an unlimited cluster creation, initially. A *distance metric*, $\wp$, must be defined, allowing a similarity measure for the clustering process. Solutions s_k generated by SM are passed to IC that attempts to group as known information, according to $\wp$. If the information is considered sufficiently new, it is kept as a center in a new cluster, c_{new}. Otherwise, redundant information activates the closest center c_i (cluster center that minimizes $\wp(s_k, c_{j=1,2,\cdots})$), causing an *assimilation* of the information represented by s_k. Considering $\mathcal{G}_{j\ (j=1,2,\cdots)}$ as all current detected clusters:

$$c_{new} = s_k \text{ if } \wp(s_k, c_j) > r_j, \forall \mathcal{G}_j, \text{ or} \tag{1}$$

$$c_i' = c_i \oplus \beta(s_k \ominus c_i), \text{ otherwise.} \tag{2}$$

where $\oplus$ and $\ominus$ are abstract operations over c_i and s_k meaning, respectively, addition and subtraction of solutions. The operation $(s_k \ominus c_i)$ means the difference between the solutions s_k and c_i, considering the distance metric. A certain percentage, β, of this difference is used to update c_i, giving c_i'. According to β, the assimilation can assume different forms: simple, crossover or path assimilations [2]. A maximum number of clusters, $\mathcal{NC}$, is defined *a priori*. To cover all the search space, a common radius r_t must be defined, working as a threshold of similarity. For the Euclidean space, for instance, r_t has been defined as [1]:

$$r_t = \frac{x_{sup} - x_{inf}}{2 \cdot \sqrt[n]{|C_t|}} \tag{3}$$

where $|C_t|$ is the current number of clusters (initially, $|C_t| = \mathcal{NC}$, in general, about 20-30 clusters), x_{sup} and x_{inf} are, respectively, the known upper and lower bounds of the domain of variable x, considering that all variables x_i have the same domain.

For combinatorial optimisation, the threshold of similarity concerns the number of moves need to change a solution in another considering a specific neighborhood relationship. For the simple 2-swap neighborhood, commonly used in permutation search spaces, r_t has been defined as [8]:

$$r_t = \lceil \lambda \cdot N \rceil \tag{4}$$

where N is the permutation size and λ is the percentage of dissimilarity, in general, about 0.9 has been set.

The assimilation process is applied over the closest center c_i, considering the new generated solution s_k. According to β in Eq. 2, the assimilation can assume different forms. The three types of assimilation, proposed in [1], are: simple, crossover and path relinking. In simple assimilation, $\beta \in [0,1]$ is a constant parameter, meaning a deterministic move of c_i in the direction of s_k. Only one internal point is generated more or less closer to c_i, depending on β, to be evaluated afterwards. The greater β, the less conservative the move is. This type of assimilation can be employed only with real-coded variables, where percentage of intervals can be applied to. Crossover assimilation means any random operation between two candidate solutions (parents), giving other ones (offsprings), similarly as a crossover operation in EAs. In this assimilation, β is an $n-$dimensional random vector and c_i' can assume a random point inside the hyper plane containing s_k and c_i. Since the whole operation is a crossover or other binary operator between s_k and c_i, it can be applied to any type of coding or even problem (combinatorial or continuous one).

Simple and crossover assimilations generate only one internal point to be evaluated afterwards. Path assimilation, instead, can generate several internal points or even external ones, holding the best evaluated one to be the new center. These exploratory moves are commonly referred in path relinking theory [9]. In this assimilation, β is a $\eta-$dimensional vector of constant and evenly spaced parameters, used to generate η samples taken in the path connecting c_i and s_k. Since each sample is evaluated by the objective function, the path assimilation itself is an intensification mechanism inside the clusters, allowing yet sampling of well-succeeded external points, extrapolating the s_k-c_i interval. For combinatorial optimisation, path assimilation has been chosen [6, 7, 8, 10], becoming the most popular of them. The more distance $\wp(c_i, s_k)$, the more potential solutions exist between c_i and s_k. The sampling process, depending on the number of instance variables, can be costly, since each solution must be evaluated by objective function.

For continuous optimisation, path relinking might be replaced by some kind of direct search movement, as Hooke-Jeeves algorithm [11], in which, the best solution found in the activated cluster is set as new center. In this case, assimilation must consider other inner points, besides c_i and s_k to proceed with the direct search algorithm.

2.3 Analyzer Module

AM component examines each cluster, indicating a probable promising cluster or a cluster to be eliminated. A *cluster density*, δ_i, is a measure that indicates the activity level inside the cluster i. For simplicity, δ_i counts the number of solutions generated by SM (selected or updated solutions in the EA case) [1, 10]. Whenever δ_i reaches a certain *threshold*, meaning that some information template becomes predominantly generated by SM, such information cluster must be better investigated to accelerate the convergence process on it. In the EA case, the promising threshold has been set according to the number of cluster, $\mathscr{NC}$, and the number of generated individuals, $|P_{new}|$, at each SM iteration. Considering that each cluster should receive the same number of individuals, a cluster i is promising whether:

$$\delta_i > \frac{\mathscr{NC}}{|P_{new}|} \tag{5}$$

On the other hand, clusters with lower δ_i are eliminated, as part of a mechanism that allows creating other centers of information, keeping framed the most active of them. At the end of each generation, the AM performs the *cooling* of all clusters, resetting the accounting of δ_i. A cluster is considered inactive when no activity has occurred in the last generation.

In EA fashion, premature convergence is an undesirable behaviour in which a population converges too early, resulting in a suboptimal solution. In this context, the individuals are concentrated in few search areas, not being able to generate offsprings that are superior to their parents. Premature convergence may happen in case of loss of diversification, i.e., every individual in the population become identical. In early ECS applications, the clustering process did not affect the set of $|P|$ solutions in SM. In this work, a control of diversification is being proposed in order to avoid premature or even undesirable convergence. By this new feature, the AM module plays another important role in the ECS. It is responsible by detecting promising areas, eliminating inactive clusters and *avoiding overcrowded areas*. Considering the same basic idea of Eq. 5, a cluster i is overcrowded whether:

$$\delta_i > \kappa \frac{\mathscr{N}\mathscr{C}}{|P_{new}|} \qquad (6)$$

where κ must be defined *a priori*, but satisfactory behavior can be reached setting it to 0.5. The population cannot be updated with a new individual grouped in a overcrowded cluster, interfering in the population diversification.

In the first experiment, an ECS, a steady-state real-coded genetic algorithm, similar to [1], was run during 10 generations (about 200 updatings), for some test functions, commonly found in literature as optimisation benchmark, as *Rastrigin* and *Langerman* [12]. Figure 2 shows the first experiment with this two test functions.

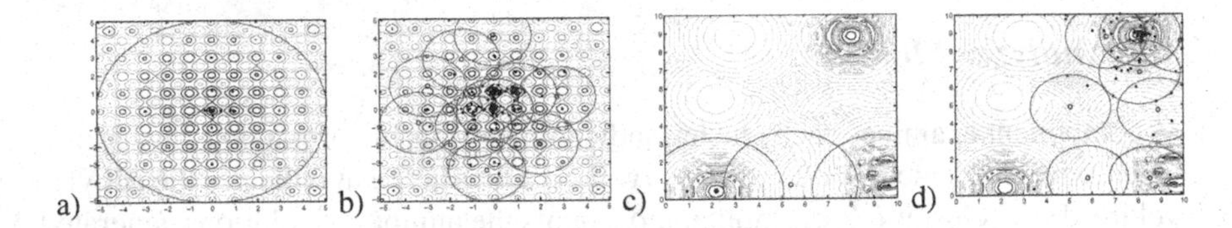

Fig. 2 Test functions after 10 generations: a) *Rastrigin* without and b) with diversification control; c) *Langerman* without and d) with diversification control

The circles show active clusters framing search areas. The dots mean individuals scattered along the promising areas. One can observe that mechanism to prevent overcrowded clusters is effective against premature convergence since ECS keeps population diversified. For example, in Figure 2a, one can note that with the same number of generations, just one cluster is sufficient to cover all population without diversification control, while several clusters are necessary to cover the population with diversification control (Figure 2b).

The second experiment consists of applying the ECS to other test-functions found in literature to analyze if the diversification control contributes to the algorithm performance improvement. The test-functions used in this experiment are showed in Table 1. A study about all of these functions can be found at [13].

Table 1 Test-functions

function	var	opt	$x_i^{low};x_i^{up}$	function	var	opt	$x_i^{low};x_i^{up}$
Ackley	n	0	-15 ; 30	Goldstein	2	3	-2 ; 2
Griewank	n	0	-600; 600	Easom	2	-1	-100;100
Schwefel	n	0	-500;500	Shekel 10	4	-10,536	-10 ; 10
Rosenbrock	n	0	-5,12;5,12	Langerman	5	-1,4	0;10
Rastrigin	n	0	-5,12;5,12	Langerman	10	-1,4	0;10
Sphere	n	0	-5,12;5,12	Michalewicz	5	-4,687	0; π
Zakharov	n	0	-5 ; 10	Michalewicz	10	-9,66	0; π
				Hartman	6	-3,322	0 ; 1

The results in Figure3 were obtained, in 20 trials, allowing ECS to perform up to 500,000 objective function calls at each trial. For each test-function, the average of well-succeeded trials (hits), the average of gap (difference between the best found solution and optimal one) and the average of objective function calls were considered to verify the algorithm performance. The parameter κ was set to $\{0.1, 0.3, 0.5, 0.7, 1.0, 1.5, 2.0\}$, respectively, where values above 1.0 allow overcrowded cluster. It is expected that very low values (e.g 0.1 and 0.3) does not allow overcrowded cluster, but the desirable convergence also is affected, causing a poor performance of the ECS.

In Figure3, one can observe the behavior of the algorithm along the experiment, varying κ. The best results concerning hits and gap were obtained for $0.5 \leq \kappa \leq 1.0$, while the best result regarding the number of function calls was obtained about $\kappa = 1.0$. One can conclude that, as expected, the diversification control can improve the ECS performance for continuous optimisation, depending on the equilibrium

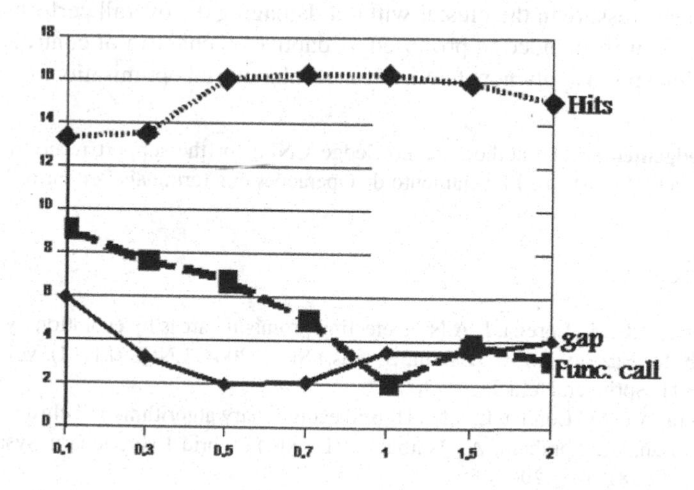

Fig. 3 Performance results: hits, function calls and GAP for test-functions

obtained during the evolutionary process. The lack of convergence prejudices the algorithm behavior.

2.4 *Local Searcher*

At last, the LS component is an internal searcher module that provides the exploitation of a supposed promising search area, framed by a cluster. This process can happen after AM having discovered a target cluster. The component LS has been implemented by a Hooke-Jeeves direct search in continuous optimisation [1]. For combinatorial optimisation, costly algorithms as VND [6] and 2-Opt heuristic, which explores several 2-Opt neighborhoods [8], have been employed. Given the pressure of density, $\mathscr{PD}$, responsable for controlling the sensibility of the component AM, the component LS has been activated, at once, if

$$\delta_i \geq \mathscr{PD} \cdot \frac{\mathscr{NS}}{|C_t|} \tag{7}$$

3 Conclusion

This paper is devoted to discuss the recent applications of Clustering Search (*CS) in combinatorial and continuous optimisation, as well as a new diversification control feature here proposed and validated. In a general way, *CS attempts to locate promising search areas by cluster of solutions. In this work, the concepts behind *CS's components are presented, taking a special care with new ways of implementing them. A new type of assimilation based on direct search algorithms is suggested and a diversification mechanism is effectively proposed and validated by performance analysis. The diversification control allowed to keep controlled the convergence pressure in the cluster without damaging the overall performance. For further work, it is intended to proposed a adaptive mechanism of control of the diversification and to apply new features for combinatorial optimisation.

Acknowledgements. The authors acknowledge CNPq for the support received in the research project " Logística e Planejamento de Operações em Terminais Portuários".

References

1. Oliveira, A.C.M., Lorena, L.A.N.: Detecting promising areas by evolutionary clustering search. In: Bazzan, A.L.C., Labidi, S. (eds.) SBIA 2004. LNCS (LNAI), vol. 3171, pp. 385–394. Springer, Heidelberg (2004)
2. Oliveira, A.C.M., Lorena, L.A.N.: Hybrid evolutionary algorithms and clustering search. In: Grosan, C., Abraham, A., Ishibuchi, H. (eds.) Hybrid Evolutionary Systems. SCI, vol. 75, pp. 81–102 (2007)

3. Chaves, A.A., Lorena, L.A.N.: Hybrid algorithms with detection of promising areas for the prize collecting travelling salesman problem. In: HIS 2005: Proceedings of the Fifth International Conference on Hybrid Intelligent Systems, pp. 49–54. IEEE Computer Society, Washington (2005)
4. Resende, M.G.C.: Greedy randomized adaptive search procedures (grasp). Journal of Global Optimization 6, 109–133 (1999)
5. Hansen, P., Mladenovic, N.: Variable neighborhood search. Computers and Operations Research 24, 1097–1100 (1997)
6. Biajoli, F.L., Lorena, L.A.N.: Clustering Search Approach for the Traveling Tournament Problem. In: Gelbukh, A., Kuri Morales, Á.F. (eds.) MICAI 2007. LNCS (LNAI), vol. 4827, pp. 83–93. Springer, Heidelberg (2007)
7. Chaves, A.A., Correa, F.A., Lorena, L.A.N.: Clustering Search Heuristic for the Capacitated p-median Problem. Springer Advances in Software Computing Series 44, 136–143 (2007)
8. Oliveira, A.C.M., Lorena, L.A.N.: Pattern Sequencing Problems by Clustering Search. In: Sichman, J.S., Coelho, H., Rezende, S.O. (eds.) IBERAMIA 2006 and SBIA 2006. LNCS (LNAI), vol. 4140, pp. 218–227. Springer, Heidelberg (2006)
9. Glover, F., Laguna, M.: Fundamentals of scatter search and path relinking. Control and Cybernetics 29(3), 653–684 (2000)
10. Filho, G.R., Nagano, M.S., Lorena, L.A.N.: Evolutionary clustering search for flowtime minimization in permutation flow shop. In: Hybrid Metaheuristics, pp. 69–81 (2007)
11. Hooke, R., Jeeves, T.A.: "Direct search" solution of numerical and statistical problems. Journal of the ACM 8(2), 212–229 (1961)
12. Digalakis, J., Margaritis, K.: An experimental study of benchmarking functions for Genetic Algorithms. IEEE Systems Transactions, 3810–3815 (2000)
13. Oliveira, A.: Algoritmos evolutivos híbridos com detecção de regiões promissoras em espaços de busca contínuos e discretos. PhD Thesis. INPE (2004)

WSAN QoS Driven Control Model for Building Operations

Alie El-Din Mady, Menouer Boubekeur, and Gregory Provan

Abstract. Currently wireless based control systems lack appropriate development methodologies and tools. The control model and its underlying wireless network are typically developed separately, which can lead to unstable and suboptimal implementations. In this paper we introduce a hybrid-based design methodology that considers the performance parameters of the Wireless Sensor and Actuator Network (WSAN) in order to develop an optimized control system tailored to the specific application environment and sensor network conditions. We first identify the boundaries of the control parameters that maintain stable and optimal control model. Within these boundaries, we determine the optimal WSAN Quality of Service (QoS) parameters through a tuning process in order to reach to optimal Control/WSAN design as illustrated in the case study. The methodology has been illustrated through a distributed lighting control developed using our hybrid/multi-agent platform.

Keywords: Hybrid System, Multi-agent System, Building Automation, WSAN, PPD-Controller.

1 Introduction

Nowadays, networked wireless devices are widely used in many applications, such as habitat monitoring, object tracking, fire detection and modern building. In particular, buildings equipped with Building Management Systems (BMS) often use a large wireless/wired sensor network. Creating distributed sensor network applications for such systems face numerous challenges in scaling, delays associated with

Alie El-Din Mady · Menouer Boubekeur · Gregory Provan
Cork Complex Systems Lab (CCSL), Computer Science Department,
University College Cork (UCC), Cork, Ireland
e-mail: {mae1,m.boubekeur,g.provan}@cs.ucc.ie

E. Corchado et al. (Eds.): SOCO 2010, AISC 73, pp. 237–247.
springerlink.com

data collection and energy consumption, which can lead to unstable systems [1], [2] (i.e., continuously oscillating around the set points), this instability might also be due to the performance tradeoffs between the control and wireless networks when designing the distributed controller. Further, the different requirements of different services place many challenges on centralized control solutions; for example, in lighting control, reaction times are anticipated within fractions of a second, whereas in HVAC control, the process dynamics is much slower and the sampling/actuation time is much larger.

Control systems and communication networks are typically designed using different platforms and principles. Control theory requires accurate, timely and lossless feedback data; however, random delays and packet loss are generally accepted in communication networks, particularly in wireless networks. Therefore, the performance of the control model relies on the network performance, due to the distribution and communication-based control. From the control perspective, the more knowledge the controller has about the system, the better the control performance can be designed to tolerate communications problems. Additional knowledge about the system can be obtained by increasing the number of sensors or sending sensor measurements more frequently. However, this increases the communication burden on the network and the network may become congested. The congestion results in longer delays and more packet losses, which degrade the control performance.

As a metric for measuring relative optimality of control performance, we have used the Mean Square Error (MSE). In terms of user comfort, this metric reflects the user's dissatisfaction in relation to the preferred set point. Moreover the degradation of the QoS at the network level may reduce user comfort; for example, a communication delay may delay reaching the optimal set point (i.e. light luminance). Second, packet losses may cause false alarms or a failure to capture real alarm data.

The objective of our work is to provide a Control/WSAN design methodology that examines the tradeoffs in optimising the building control in relation to user comfort, safety and reliability. These factors are dependent on optimal control parameters and enhanced WSAN QoS [3]. As shown in Fig. 1, we start by identifying the boundaries of the control parameters that maintain stable and optimal control model. Within these boundaries, we determine the optimal WSAN QoS parameters through a tuning process to reach to optimal Control/WSAN design as illustrated in the case study.

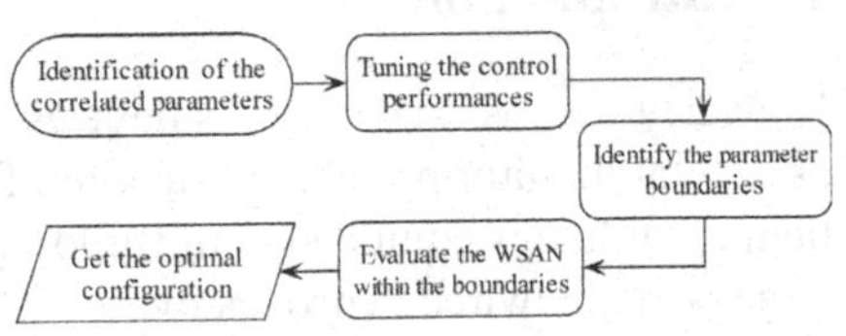

Fig. 1 Design Methodology

Our research extends prior work in the area, e.g., [4], by exploring the impact of the control performance on the WSAN and vice versa. [4] provides a cross layer methodology to link the standard design layers of an Open System Interconnection (OSI). This methodology ignores the performance of the WSAN and fails to consider linking the performance evaluation of the different layers which may improve control performance but degrade that of the other layers. We have selected the Medium Access Control (MAC) protocol and the Link technique design; we do not consider the network layer because the underlying example uses a point-to-point linking technique. The impact of changing the correlated parameters on both control performance and the WSAN QoS has been considered, with priority given to the objectives of the application, as represented by the control requirements.

Two tuning phases have been considered in the proposed methodology. The first considers tuning control performance to get the best correlated parameter values; for this we calculate the parameter variation boundaries. The second one deals with the WSAN QoS; for this we explore the search population within the boundaries provided previously, to determine the optimal Control/WSAN configuration.

The remainder of the paper is organized as follows: Section 2 introduces a new control strategy, called the Parameterised Predictable Distributed (PPD) control strategy, its WSAN model and the scenario specification considered. In Section 3, the hybrid/multi-agent model for the PPD-Controller is explained. The refined approach for the Control/WSAN-correlated parameters is provided in Section 4. The experimental results for our case study are provided in Section 5 and finally Section 6 highlights our conclusions and plans for future work.

2 Parameterizable/Predictable Distributed Controller

This section introduces our new Parameterizable and Predictable Distributed controller (PPD-Controller) for automated lighting systems. The PPD-Controller offers a distributed solution and aims to increase the control reliability, scalability, resource sharing and concurrency. In this section, we briefly describe the scenario specification, the control strategy and the WSAN model.

2.1 Scenario Specification

We consider an open office area with typical architecture, as shown in Fig. 2. It contains 10 controlled zones; each zone contains one artificial light, one light sensor and one Radio-Frequency Identification (RFID) receiver. Therc are 4 windows/bindings on the right and left boarders of the open area and a fix number of predefined person positions.

For the lighting model we consider integrating blinding and lighting controls. In order to enhance the efficiency of the resulting control model, an optimization technique has been implemented as it selects the light luminance and blind position depending on the user preferences and the power consumed due to the artificial light and the blinding actuators.

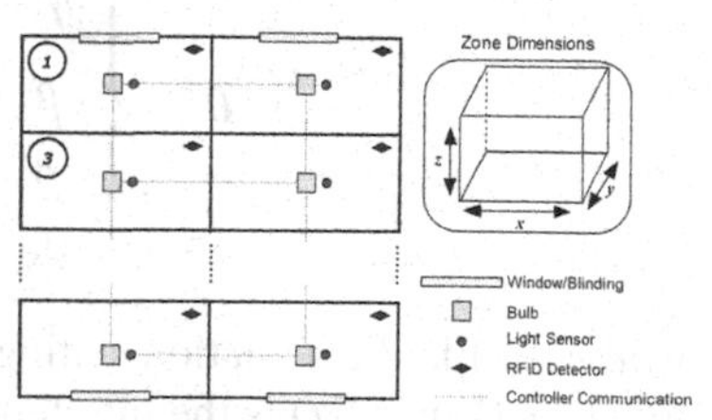

Fig. 2 Model Specification

As a summary, the lighting control scenario behaves as follows:

1. The user can switch on/off the automatic lighting system for several zones, or for all the system through a technician.
2. The users provide their preferences (light luminance and blinding position).
3. A person is tracked in each zone using for example RFID, his preferences are ignored whenever he leaves his zone.
4. A local optimization engine receives the user preferences and sends back the optimal settings.
5. The controller controls the artificial light and the blinding actuators in order to reach the user preferences considering the daylight luminance and the light interferences coming from the adjacent zones.

2.2 Control Strategy Description

Fig. 3 shows the model of a local controller and its interactions with the environments. The preference solver receives the user preferences for each zone, sends the optimal light luminance and blinding position back to the optimization engine. This latter uses Genetic Algorithm/Simulated Annealing (GASA) algorithm [5] in order to calculate the optimal actuation settings, it sends them back to the PI-Controller. The PI-Controller, is used to predict the next actuation setting for the lighting level in a close loop fashion [6] using Eq. 1. It actuates the artificial light and the blinding position according to the optimum settings. Whenever preferences change, the optimization step is updated; otherwise, the PI-Controller actuates relying on the external light and the light interference. The Light/Blinding Occlusion Preference Solver agent is used to provide the intermediate solution between several luminance/glare preferences in the same controlled zone. It applies a Low Pass Filter (LPF) in order to prevent exceeding a predefined threshold (700 Lux for luminance and 100% for the blinding position). The control equations are given by:

$$\begin{aligned} A(t+1) &= A(t) + \theta \\ U(t) &= A(t) + E(t) + I(t) \end{aligned} \tag{1}$$

$$\theta = \begin{cases} \gamma - \dfrac{\beta}{\rho}, & \forall\, U(t) - S(t) > \varepsilon \\ \dfrac{\beta}{\rho} - \gamma, & \forall\, S(t) - U(t) > \varepsilon \\ 0, & \forall\, |S(t) - U(t)| \leq \varepsilon, \end{cases}$$

where $A(t)$ is the actuation setting for light/blinding actuators, $E(t)$ is the daylight intensity (Lux), $I(t)$ is the interference light intensity (Lux), $U(t)$ is the sensed light intensity (Lux), $S(t)$ is the optimal preference settings, ε is the luminance level produced from a single dimming level (70 Lux), β is the maximum light intensity error (700 Lux), γ is the minimal light intensity error (0 Lux) and ρ is the total number of dimming levels (10 levels).

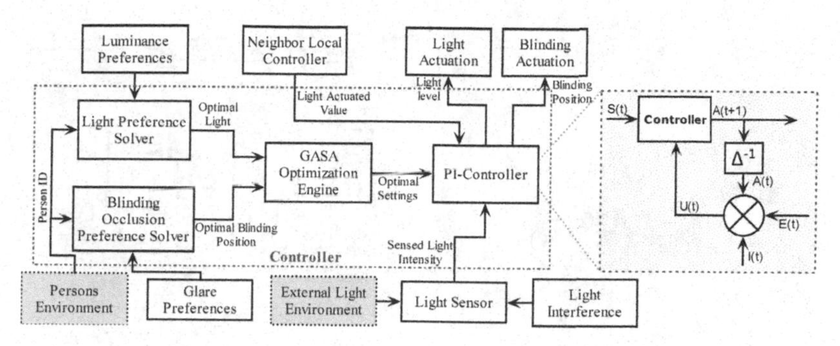

Fig. 3 Control Model

2.3 WSAN Modelling for the PPD-Controller

In order to evaluate the WSAN performance for the PPD-Controller, we have modelled the WSAN using the VisualSense tool [7]. We have also considered the Tyndall [8] sensor node as a reference for the model parameters. The Time Division Multiple Access (TDMA-based) MAC protocol [9] is used in the contention-free period, which leads to a free collision probability. Fig. 4(a) shows the WSAN model used for evaluating 4 zones (1, 3, 4, 5) (Fig. 2). The PPD-Controller in zone 3 has been selected to be evaluated as it constitutes the bottleneck in the model, since it is the most heavily used due to its communication with the other 3 controllers (1, 4, 5), their RFIDs and sensors. In relation to the WSAN performance, the following QoS metrics [10] have been identified: $Channel\ Throughput$, $Controller\ Duty\ Cycle$, $Buffer\ Size$, $Response\ Time(Delay)$, and $Sensorbattery\ LifeTime$.

When modelling the WSAN for PPD-Controller, we distinguished four models:

Communication channels model: 2 channels are considered for the wireless communication, one channel for light sensors and the local controllers (Zigbee band, i.e. 2.4 GHz) and other for the RFIDs (RF band, i.e., 324 MHz). The power propagation factor in the communication channels is $\frac{1}{4\Pi r^2}$, where r is the distance between the transmitter and the receiver, and the loss probability in each channel is 2%.

Light sensor model: The sensor sends the Lux measured value and the sensor ID to the controller using a fixed sampling rate and frequency offset, as shown in Fig. 4(c). The sensor coverage area is 3 meters (distributed in sphere area) and its power transmission is 0.1 $watt/m^2$. In order to show the effect of the battery discharging on the sensor transmission range, we have assumed that the range is decreasing by 0.1 meter each event that follows Poisson distribution with mean time equals to 20 times the sensor sampling rate.

RFID model: The RFID detection range is 1.5 meter and its power transmission is 0.1 $watt/m^2$. As shown in Fig. 4(d), the RFID sends its ID with a fixed sampling rate and frequency offset. Moreover, the movement of the RFID is modelled as a sin wave sampled every 0.3 minute.

Controller/Receiver model: In this model, shown in Fig. 4(b), we have considered the received packets number, buffer size and the controller duty cycle. However, the

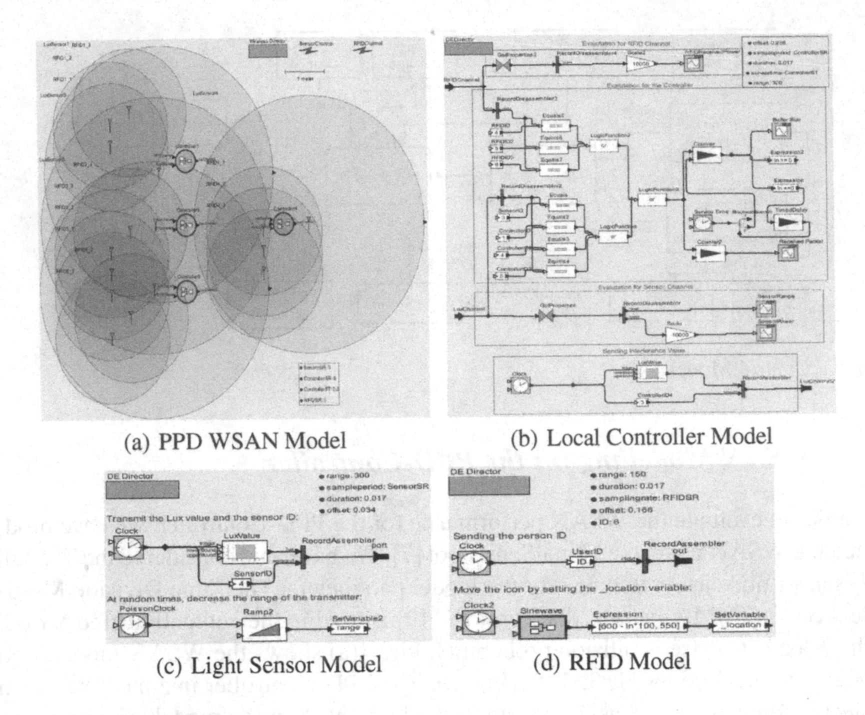

(a) PPD WSAN Model (b) Local Controller Model

(c) Light Sensor Model (d) RFID Model

Fig. 4 WSAN Model

controller service time is fixed per received packet. The communication between the neighbouring controllers also uses the sensor channel.

3 Design and Model for PPD-Controller Using Hybrid Platform

In order to simulate the lighting system and evaluate its performance, the system and its environment have been modelled using the Charon toolset [11]. The Charon toolset provides a platform to create a hybrid/multi-agent system, using a hierarchy framework. Based on Charon's use of agents to specify control entities, in the PPD-Controller, we use an agent for the global controller, which sends configuration parameters to the local controllers. Two other agents model the environments (external light and presence). For each zone, 4 agents are used: RFID, light sensor, blinding controller and local controller. As mentioned earlier, the global controller sets the configuration parameters for the local controllers, e.g., it activates/deactivates some controllers (e.g. blinding controller) or some functions inside a controller (e.g. manual or automatic). The local controller contains 2 subagents, one of which is used to receive and calculate the light interferences coming from the adjacent zones. The agent also predicts light interferences using Linear Prediction Coding (LPC) algorithm [12], based on the preferences history when starting of the scheduling flow. The other agent is used to send the actuation values and trigger the optimization engine. Each agent contains the modes describing its behaviour. There are two main environments for the lighting system, the daylight and the person movement environments. In order to verify the behaviour of the PPD-Controller, both environments have been modelled using hybrid systems, as the daylight model has continuous behaviour while the presence model has discrete behaviour. In the daylight model, five periods have been modelled as a first order differential equation with a constant slope (using linear hybrid automata [13]). During the first and last four hours of the day, the daylight slope and luminance are equal to zero, while during the second four hours the slope is equal to 100, which means that the maximum intensity in the day is 4000 Lux. In the next eight hours the slope is equal to zero and then goes to -100 in the following four hours, in order to reach zero luminance again at the end of the day. The light intensity that comes to the controlled zone is a percentage of the daylight intensity; this percentage relies on the dimensions of the window. In this model, 8% of the daylight is considered as the external light coming into the controlled zone [6].

We model occupant movements in the controlled zone using a deterministic distribution with respect to the time of day. In the first and last seven hours of the day, no one is in the zone; from 7:00 to 10:00 AM people arrive successively; during the next seven hours occupants enter or exit with a 50% probability; and finally, over the next two hours people leave individually.

4 Control/WSAN Refinement Approach

As stated earlier, in modern buildings, distributed controllers over large wireless/wired sensor/actuator network face the challenge of achieving good WSAN performances while designing the control application. The case where both control and WSAN models are designed separately may lead to unstable and suboptimal implementations. In this research work we assume a high correlation between the performance parameters of both control and WSAN models. For example, if the WSAN has received many requests at a certain moment, this will lead to either delay in responding to the next request (in order to serve all the buffered requests) or dropping some requests which will create unexpected behaviour in the environment. In this section we explain our approach for an integrated design of both control and WSAN.

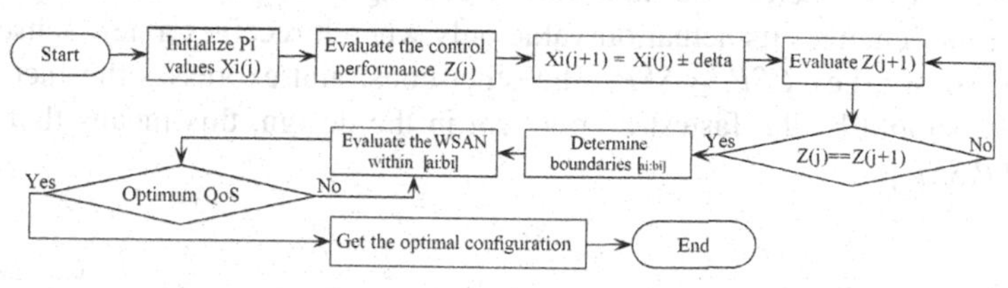

Fig. 5 WSAN/Control Evaluation Algorithm

Fig. 5 shows the flowchart of the approach:

1. We start by identifying the correlated parameters P_i which mutually affect both WSAN QoS and the control performance.
2. Initialize the P_i with acceptable values $P_i(0)$.
3. With the assumption that the control performance has higher priority, evaluate the control performance (MSE) according to the initial parameters.
4. Increase/decrease P_i with a (delta) step using Gradient Descent algorithm and remove the inconsistent solutions, as will be explained later.
5. Evaluate the MSE according to $P_i(j)$, which indicates the value for P_i at instance j.
6. Repeat step 5 untill getting the acceptable control performance, and hence identify the boundaries $[a_i, b_i]$ for each parameter P_i.
7. Evaluate the QoS of the WSAN within the identified boundaries $[a_i, b_i]$.
8. Repeat step 7 untill the QoS equals a predefined criteria for WSAN.

Through studying the correlated parameter space of the PPD-Controller/WSAN, we have identified that the Sensor Sampling Period (*SSP*), Controller Sampling Period (*CSP*) and Controller Service Time (*CST*) are the correlated parameters $P_i \subseteq \{SSP, CSP, CST\}$. However, other parameters may affect the WSAN or the control separately; for example, the sampling period for the RFID affects the WSAN QoS but it does not affect the controller. As it is handled by the controller in an event-based model, the controller considers only the occupant presence and not the frequency of the sampling period. Relying on the aforementioned P_i, we can conclude that $P_i(j)$ depends mainly on the control strategy, as in the centralized control model, $P_i(j)$ will have different values than the PPD control model (i.e high *SSP* and *CSP*, and low *CST*).

5 Experimental Results

The proposed methodology has been used to design the PPD-Controller presented in Section 2 and its underlying WSAN model for the zones 1, 3, 4 and 5. We used a Gradient Descent algorithm to identify the direction for improvement. In order to optimize the solution space, we defined two design constraints to determine $P_i(j)$ and eliminate the inconsistent combinations that do not match these constraints. The first constraint considers that *CSP* is used to exchange the actuation values, the controller can then detect the interference coming from other zones. In this case, the controller changes its actuation value only when it receives a new sensed value from the sensor, i.e., $CSP \geq SSP$. The other constraint expresses the fact that the controller should be the fastest component in the design, this means that $CST < \min\{CSP, SSP\}$.

5.1 Control Refinement

The main metric used for evaluating the control performance is the MSE, calculated using Eq. 2. The MSE indicates user dissatisfaction, i.e., it indicates how the actuated values deviate from the preferred ones. As a starting configuration, for P_i we have chosen $P_i(0) = \{15min, 15min, 0.5min\}$, which corresponds to a typical system settings. The Gradient Descent algorithm is then used in order to identify the next values for P_i.

As shown in Table 1, when $P_i(1) = \{20min, 20min, 0.5min\}$, which indicates an increase in *SSP* and *CSP*, the controller performance degraded. However when *SSP* and *CSP* are decreased, $P_i(1) = \{10min, 10min, 0.5min\}$, the control performance improved. Therefore, the improvement is achieved by decreasing the initial value. Accordingly, the search population considered for control performance evaluation is $P_1(j) = P_2(j) = \{15min, 10min, 5min, 0.6min\}$ and $P_3(j) = \{0.5min\}$. Note that $P_1(j)$ and $P_2(j)$ are stopped at 0.6 min according to the pre-defined constraint stipulating that the controller is the fastest system component ($P_3(j)$). At the start *SSP* is equal to *CSP* according to the pre-defined constraint, $CSP \geq SSP$, we explore then the search space while evaluating the MSE to identify the optimal point corresponding to the Minimum MSE (MMSE).

After evaluating the control performance for all the search population identified previously, we found out that the MMSE is at $P_i(j) = \{0.6min, 5min, 0.5min\}$; however, when $P_i(j) = \{0.6min, 10min, 0.5min\}$, the performance is improved over the previous evaluated point, and hence the controller's optimal point is $P_i(j) = \{0.6min, 5min, 0.5min\}$. Accordingly, the boundaries for *SSP* is $[a_1, b_1] =]0.5, 0.6]$, *CSP* is $[a_2, b_2] = [0.6, 5]$ and *CST* is $[a_3, b_3] = [0, 0.5]$.

$$MSE = \frac{\sum_{a=1}^{N} \frac{\sum_{k=1}^{M} (U_a(k) - S_a(k))^2}{M}}{N}, \tag{2}$$

where N is the total number of zones, M is the total number of samples.

Table 1 Controller Performance

$P_i(j)$	MSE (Lux)
{20, 20, 0.5}	52.53
{15, 15, 0.5}	40.13
{10, 10, 0.5}	39.17
{5, 5, 0.5}	36.48
{0.6, 0.6, 0.5}	35.64
{0.6, 5, 0.5}	34.32
{0.6, 10, 0.5}	45.89

5.2 *WSAN Refinement*

When studying the WSAN, assuming that *SSP* and *CSP* are fast enough, the most effective QoS metric for the user comfort is *Response Time*, as it reflects how much time is needed to serve an update detected by the light sensor or the neighbourhood controller. It appears that the *CST* is not affected by the WSAN QoS, since it is linked with the MAC layer switching, which implies timing constraints. In exploring the impact of the *CST* on the WSAN QoS, we selected the stopping WSAN criteria based on the *Response Time* metric. Assuming that the required criterion for the WSAN evaluation is $Response\ Time = CST$, we modify the *CST* within the boundaries obtained at the control refinement stage, as shown in Table 2. The table shows the search space and the corresponding WSAN QoS metrics, including Channel Throughput, Controller Duty Cycle, Buffer Size, Response Time and Battery Life Time.

Table 2 WSAN QoS

$P_i(j)$	Channel Throughput	Controller Duty Cycle	Buffer Size	Response Time	Battery Life Time
$\{5, 5, 0.5\}$	1.4 packet/min	70%	7 packets	1 min	79.64 days
$\{0.6, 5, 0.5\}$	5.67 packet/min	100%	367 packets	183.5 min	76.47 days
$\{0.6, 5, 0.4\}$	5.69 packet/min	100%	319 packets	127.6 min	76.47 days
$\{0.6, 5, 0.3\}$	4.63 packet/min	100%	148 packets	39 min	76.47 days
$\{0.6, 5, 0.2\}$	5.5 packet/min	100%	65 packets	9.4 min	76.47 days
$\{0.6, 5, 0.1\}$	5.73 packet/min	57.3%	5 packets	0.1 min	76.47 days

In relation to the sampling period (*SSP*, *CSP*), it is obvious that the slower the period, the better is WSAN QoS. We have thus chosen the higher values from the control (*SSP*, *CSP*) boundaries ($SSP = 0.6$, $CSP = 5$). We can conclude that the optimal point matching the Control/WSAN requirements is $P_i(j) = \{0.6, 5, 0.1\}$ and moreover it shows a good improvement in the *Control Duty Cycle* and the *Buffer Size* metrics. However the *Battery Life Time* and *Channel Throughput* have almost the same effect. It is obvious that the selection of the design points presents a tradeoff between battery life and user comfort (reflected by sampling period). If we consider increasing the *SSP* to 5 min, we should expect $Response\ Time = 2 \times CST$ and a slightly worse control performance, as depicted in Table 1.

6 Conclusion

In this article, we have provided within our hybrid/multi-agent platform a refinement methodology for improving the Control/WSAN performance within the building automation domain. Such an improvement plays a key role in guaranteeing properties such as safety, accuracy, stability and reactiveness, which greatly impact user comfort. The developed methodology can configure the Control/WSAN-correlated parameters, and thereby reach an efficient configuration. The approach has been tested on an PPD-Controller used for lighting systems and the impact of changing the correlated parameters on both control performance and the WSAN QoS has been considered. At this stage, we prioritise the objectives of the application, as represented in the control requirements.

As future work, we intend to apply our methodology to Heating, Ventilating, and Air Conditioning (HVAC) system, as this presents more interesting challenges in relation to user comfort and control stability. We also aim to deploy a demonstration of the developed system in the Environmental Research Institute (ERI) building, which is the ITOBO Living Laboratory [14]. The benefit of cross-layer modelling for distributed control constitutes an important research topic that we also intend to pursue in future work.

Acknowledgements. This work was funded by Science Foundation Ireland (SFI) grant 06-SRC-I1091.

References

1. Nakamura, M., Sakurai, A., Furubo, S., Ban, H.: Collaborative processing in Mote-based sensor/actuator networks for environment control application. In: Signal Processing, Elsevier, Amsterdam (2008)
2. Lin, Y., Megerian, S.: Low cost distributed actuation in largescale Ad hoc sensor-actuator networks. In: International Conference on Wireless Networks, Communications and Mobile Computing, pp. 975–980 (2005)
3. Akyildiz, I.F., Kasimoglu, I.H.: Wireless sensor and actor networks: research challenges. In: Ad hoc networks, Elsevier, Amsterdam (2004)
4. Liu, X., Goldsmith, A.: Cross-layer Design of Distributed Control over Wireless Networks. In: Advances in Control, Communications Networks (2005)
5. El-Hosseini, M.A., Hassanien, A.E., Abraham, A., Al-Qaheri, H.: Genetic Annealing Optimization: Design and Real World Applications. In: Proc. of Eighth International Conference on Intelligent Systems Design and Applications, ISDA 2008 (2008)
6. Kolokotsa, D., Pouliezos, A., Stavrakakis, G., Lazos, C.: Predictive control techniques for energy and indoor environmental quality management in buildings. In: Building and Environment (2008)
7. The Ptolemy Project, `http://ptolemy.eecs.berkeley.edu/`
8. Tyndall National Institute, `http://www.tyndall.ie/`
9. Liu, A., Yu, H., Li, L.: An energy-efficiency and collision-free MAC protocol for wireless sensor networks. In: 61st IEEE Vehicular Technology Conference (2005)
10. Demirkol, I., Ersoy, C., Alagoz, F.: MAC protocols for wireless sensor networks: a survey. IEEE Communications Magazine (2006)
11. Charon: Modular Specification of Hybrid Systems, `http://rtg.cis.upenn.edu/mobies/charon/`
12. Garg, M.: Linear Prediction Algorithms. Indian Institute of technology, Bombay, India (2003)
13. Henzinger, T. A.: The theory of hybrid automata. In: Proc. 11th Annual IEEE Symposium on Logic in Computer Science (LICS 1996), pp. 278–292 (1996)
14. Environmental Research Institute, `http://www.ucc.ie/en/ERI/`

Intelligent Hybrid Control Model for Lighting Systems Using Constraint-Based Optimisation

Alie El-Din Mady, Menouer Boubekeur, Gregory Provan,
Conor Ryan, and Kenneth N. Brown

Abstract. Lighting systems consume a considerable proportion of total energy budgets, particularly for retail and public-office applications, and hence their optimisation can save considerable amounts of energy. This paper proposes an intelligent control strategy to operate the office luminance in order to enhance user comfort and reduce energy consumption. The strategy is applied to an open office scenario, where the controller and the environments are modelled using a hybrid/multi-agent platform. The developed controller uses a constraint-based optimisation technique to compute the optimal settings. We describe the different modelling steps, including the optimisation technique, and outline the simulation results and potential energy benefits of the proposed controller.

Keywords: Hybrid System, Multi-agent System, Building Automation, Constraint-Based Optimisation.

1 Introduction

An intelligent building incorporates a Building Management System (BMS) which aims to optimise energy usage while trying to optimise user comfort. One major energy consumer in buildings is lighting, which can account for up to 30% of total energy waste in some retail and public offices [1]. This energy inefficiency is due to a lack of energy-efficient lighting controllers.

Alie El-Din Mady · Menouer Boubekeur · Gregory Provan
Cork Complex Systems Lab (CCSL), Computer Science Department,
University College Cork (UCC), Cork, Ireland
e-mail: {mae1,m.boubekeur,g.provan}@cs.ucc.ie

Conor Ryan · Kenneth N. Brown
Cork Constraint Computation Centre (4C), Computer Science Department,
University College Cork (UCC), Cork, Ireland
e-mail: {c.ryan,k.brown}@4c.ucc.ie

E. Corchado et al. (Eds.): SOCO 2010, AISC 73, pp. 249–259.
springerlink.com

The aim of our research is to define a methodology for efficient modelling and integration of building management system services, in particular lighting and Heating, Ventilating, and Air Conditioning (HVAC) systems. We assume that building automation models can be represented using hybrid systems models [2], since hybrid systems can represent both the discrete-valued and continuous differential-equation-based relations essential for such models.

In this paper we show how we can use component-based hybrid systems to model and simulate an intelligent lighting controller. The developed lighting system tracks the presence of people in the different controlled areas and allows users to express preferences for interior lighting levels. The control system accommodates such preferences for all occupants within a zone, by optimising a global preference/energy function using a constraint-based optimiser to compute the optimal light luminance levels specified by the user preferences, and the power consumed by the artificial light and the blinding actuators. The centralized controller then maintains this set-point by adjusting window blinds to control the exterior light levels, and by dimming the interior lights.

The remainder of the paper is organized as follows: Section 2 introduces our modelling platform and describes the system scenario specification, the control strategy and the corresponding hybrid modelling for the control and its environments. The constraint-base optimisation technique is explained in Section 3. Section 4 outlines the simulation results. Finally, we conclude the paper in Section 5.

2 Hybrid Modelling for Intelligent Lighting System

We now apply our design methodology to develop an intelligent controller for a lighting system. In this section we introduce our modeling framework, describe the scenario specification for the lighting system model, the control strategy and finally the corresponding hybrid modelling.

2.1 Hybrid Platform for Building Control

Building systems are a perfect example of hybrid systems, where continuous and discrete dynamics must be used for modelling. For example, heat dissipation and luminosity follow a continuous dynamics whereas presence detection is a discrete nature. In our work we show how we can use a component-based hybrid-systems modelling framework to generate models for simulation and verification.

To implement the hybrid systems for building models, we using the CHARON tool [3]. We assume that we can create/redesign a system-level model by composing components from a component library [[4], [5]]. We call a well-defined model fragment a component. We assume that each component can operate in a set of behaviour-modes, where a mode *M* denotes the state in which the component is operating. For example, a pump component can take on modes nominal, high-output, blocked and activating.

Our ongoing research work consists of developing an integrated platform for intelligent control of building automation systems. This platform provides, among other features, predictability, reconfiguration, distribution and building energy optimisation.

The system design flow starts by defining relevant scenarios to be operated within the building. These scenarios are defined using the Unified Modelling Language (UML) [6]. The UML models are interpreted using specific models for simulations and analysis purposes. At this level we allow an optimisation loop to optimise the model at an early stage of the development. When the simulation gives satisfactory results, the models are auto-translated into embeddable code to be deployed over a distributed sensor/actuator network [7]. The integration process is performed through the implementation of a model-/service-based middleware [8] platform allowing components connection and data exchange. All the different components of the architecture collaborate with the requirements module.

The main features of our platform will be illustrated through an example of a lighting system for an office area. This example illustrates the combination of discrete-event behaviour (presence detection, light actuation levels) and hybrid properties for the luminosity control, i.e., where both discrete and continuous aspects are considered.

2.2 Scenario Specification

We have adopted the architecture shown in Fig. 1(a) for our work. We focus on an open office area, which contains 6 controlled zones, where each zone contains one artificial light and one light sensor. One Radio-Frequency Identification (RFID) receiver is used to cover the whole area; there is one window/binding in the left boarder of the conceded area and a fix number of predefined person positions. For the lighting model we integrate blinding and lighting controls. In order to enhance the efficiency of the resulting control model, an optimisation technique has been implemented, as explained in Section 3.

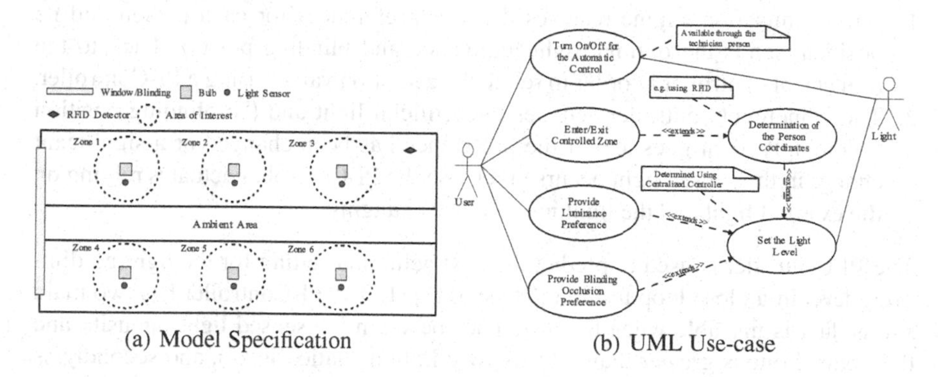

(a) Model Specification (b) UML Use-case

Fig. 1 Lighting System Specification

As a summary, the lighting control scenario, as described in Fig. 1(b), behaves as follows:

1. The user can switch on/off the automatic lighting system for several zones, or for all the system through a technician.
2. The users provide their light luminance preferences.
3. A person is tracked in each zone using RFID, and his preferences are considered wherever he is located.
4. An optimisation engine receives the user preferences and sends back the optimal settings.
5. The controller controls the artificial light and the blinding actuators in order to reach the user preferences considering the daylight luminance and the light interferences coming from the adjacent zones.

2.3 *Control Strategy Modelling*

Fig. 2 shows the agents of the control model and its interactions with the environment agents. The controller follows the following scenario in order to control the light intensity:

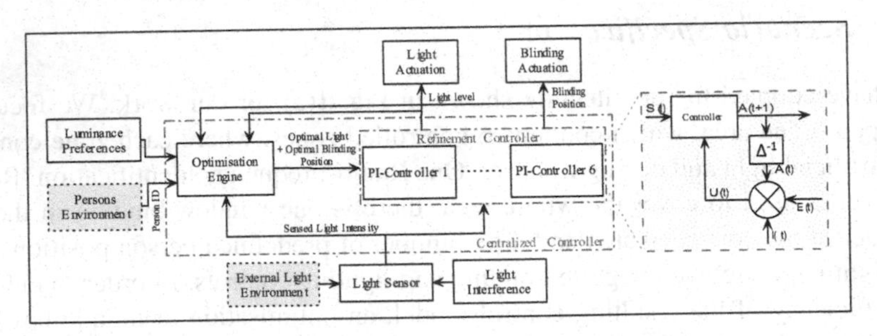

Fig. 2 Control Model

1. The optimisation engine receives the user preferences for each person and its position, sends the optimal light luminance and blinding position back to the refinement controller in order to refine the actuation values using a PI-Controller.
2. The refinement controller actuates the artificial light and the blinding position accordingly, then goes to 1 if the preference has been changed or a significant change in the sensed light occurs otherwise the PI-Controller actuates relying on the external light and the light interference changing.

The PI-Controller is used to predict the next actuation setting for the lighting dimming level in a close loop fashion [9] using Eq. 1. The PI-Controller has two main status, first is unstable when the difference between the sensed light intensity and the optimal one is greater than 70 Lux (one light actuation level), and secondly, is stable, if the difference is less than or equal to 70 Lux.

$$A(t+1) = A(t) + \alpha \tag{1}$$
$$U(t) = A(t) + E(t) + I(t)$$

$$\alpha = \begin{cases} \lfloor \frac{U(t)-S(t)}{\varepsilon} \rceil \times -\varepsilon, & \forall\, U(t)-S(t) > \varepsilon \\ \lfloor \frac{S(t)-U(t)}{\varepsilon} \rceil \times \varepsilon, & \forall\, S(t)-U(t) > \varepsilon \\ 0, & \forall\, |S(t)-U(t)| \leq \varepsilon \end{cases}$$

Where, $A(t)$ is the actuation setting for light/blinding actuators, $E(t)$ is the daylight intensity (Lux), $I(t)$ is the interference light intensity (Lux), $U(t)$ is the sensed light intensity (Lux), $S(t)$ is the optimal preference settings and ε is the luminancc level produced from a single dimming level (70 Lux).

2.4 Hybrid Modelling

In the Charon modelling shown in Fig. 2, two types of agent have been used to model the control- and environment agents.

Control Agents : One main agent is used for the refinement controller, such that one subagent is used to refine the actuation values in each zone using a PI-Controller as depicted in Fig. 3(c). Another agent is used to call the optimisation engine; it follows the behavioral mode depicted in Fig. 3(b). This agent is triggered whenever the user preferences change or a significant change in the sensed light is observed. Finally, the sensor agent has been modelled as shown in Fig.3(a), as it updates the internal light value every sampling period, based on the actuation value, the light interference and the daylight light coming to the sensor. It considers an intensity attenuation factor of $1/r^2$, where r is the distance from the light source to the sensor.

Three environment agents have been used to verify the control behaviour as following:

Person Movements : One agent is used to model a person's movements. This agent uses a Markov chain to model the person presence in the zones using a Phase-type Distribution [10]. As shown in Fig. 4, λ is the transition probability between each zone and the ambient area. The time consumed (t) in each zone follows Eq. 2. Fig.3(e) shows the hierarchal hybrid automata for the Markov chain model. When a person moves from his zone (current zone) to the next zone, the model goes to the deeper level in order to reflect his transition to the other zones. If the user moves to the absorption state, that means he goes out of the controlled area.

$$f(t,\lambda) = 1 - e^{-\lambda t} \tag{2}$$

Daylight Intensity : In the daylight model shown in Fig. 3(d), five periods have been modelled as a first order differential equation with a constant slope using linear hybrid automata [11]. During the first and last four hours of the day, the daylight slope and luminance are equal to zero, while during the second four hours the slope is equal to 100 which means that the maximum intensity in the day is 4000 Lux. In the next eight hours the slope is equal to zero and then goes to -100 in the following four hours in order to reach zero luminance again at the end of the day. The light intensity that comes to the controlled zone is a percentage of the daylight intensity; this percentage relies on the dimensions of the window. In this model, 8% of the daylight is considered as the external light coming into the controlled zone [9].

Window Blinding Occlusion : One Charon agent is used to model the blinding occlusion as an algebraic equation. It calculates the daylight percentage coming to the controlled zone as a linear proportion from the blinding position as following

```
alge{ExternalIndoorLight ==
        ExternalLight*(BlindAct/TotalControlLevels)}
```

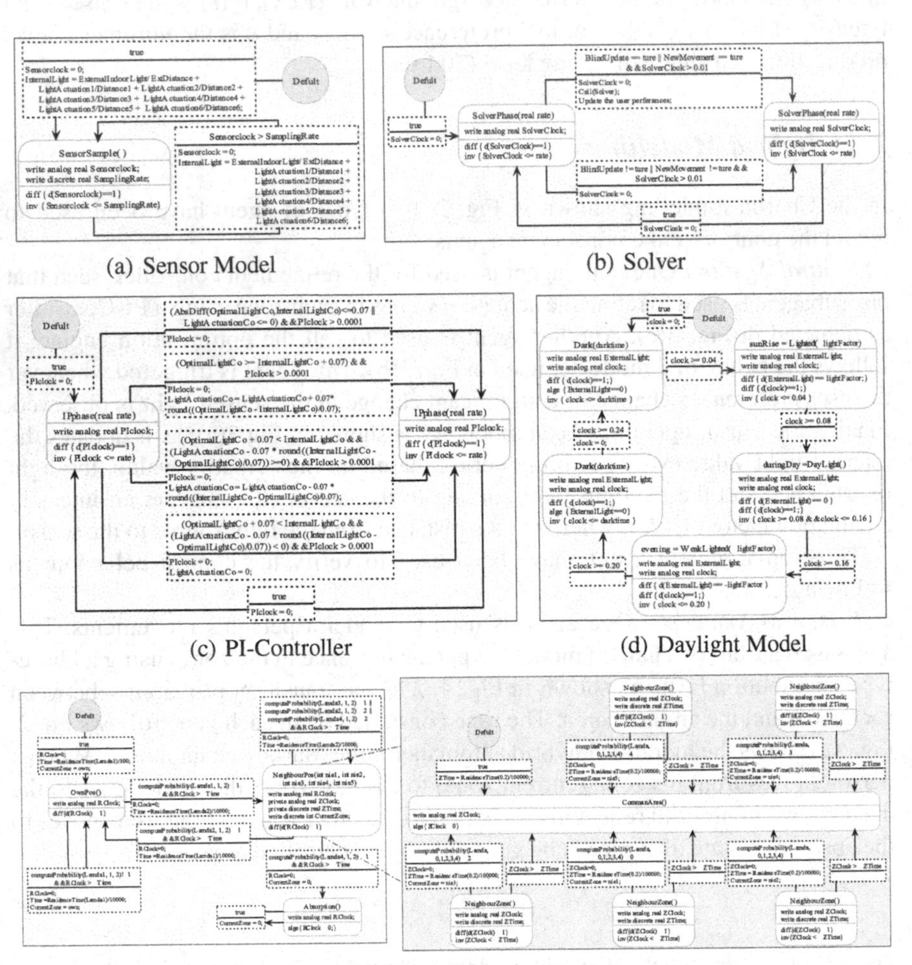

Fig. 3 Linear Hybrid Automata Models

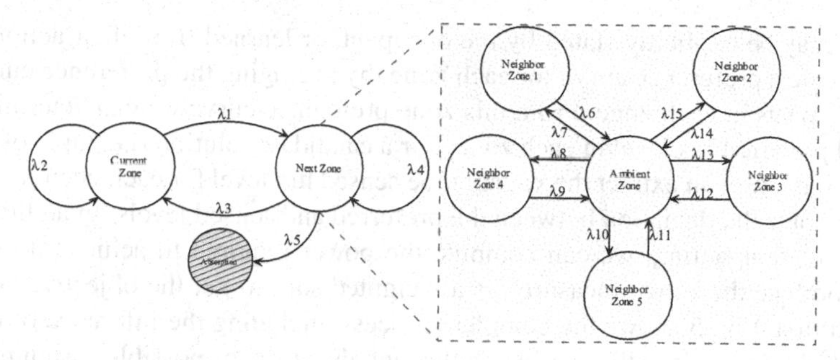

Fig. 4 Markov Chain for Person Movements

3 Constraint-Based Optimisation

The purpose of the optimisation engine is to compute settings for the lights which optimise both the energy use and the occupant satisfaction levels. The computation is based on the inferred external sunlight and the stated occupant preferences, and uses an idealised model of the domain. The derived settings together with the preferences are sent to the controller as initial actuator settings. The controller then controls the lighting around these initial values. When the occupants change, or the actuated levels diverge significantly from the initial values, the problem is re-specified to the optimisation engine, which re-computes and issues new settings.

We model the building environment, the actuated lights and their effect on the environment as a Constraint Optimisation Problem. A constraint problem [12] consists of a set of variables, a domain of possible values for each variable, a set of constraints over the assignment of values to variable which restrict the values that may be assigned simultaneously, and an objective function over the assignments. A solution is an assignment of one value to each variable, such that no constraint is violated. An optimal solution is one with the highest objective value. Solutions may be obtained by any suitable method, including backtracking and logical reasoning, mathematical programming, or local search.

The variables in our model include the setpoints of the actuated lights, the blinding level, the lux levels at the zone sensors, and the lux level of the external light. The domains for the light setpoints are the 11 integer values from 0 (off) to 10 (full power) and the domain for the blinding is the 5 integer values from 0 (fully closed) to 4 (fully open). The constraints relate the actuation values to the lux levels (and so in our model the setpoints are decision variables, the external light is a constant, and the sensed lux levels are dependent variables). We assume that each light source contributes a fraction of its luminance to each zone, using the same underlying model in the simulator. A solution is then a set of actuation points and lux levels such that the constraints are satisfied. When the optimisation model is activated, the controller passes it the actuation values and the sensed lux levels for the current state; from this, we can infer the current value of external light.

We associate with each individual occupant a preference curve, which associates a degree of satisfaction from 0 to 1 with each possible lux level. These preference curves may be explicitly stated by the occupant, or learned from their actions. We then create a preference curve for each zone, by averaging the preference curves of the occupants in each zone. From this zone preference curve we can determine the overall preferred lux level in each zone. For a candidate solution (i.e. tuple of actuation points), we can extract the value of the sensed lux level for each occupied zone, and compute the deviation between the preferred and sensed levels. In addition, for each actuation setting, we can compute the power required to achieve it, and we then combine these two measures as a weighted sum to get the objective value of the solution. Fig. 5 shows the complete process, including the inference of the external light. Our aim is then to search through the space of possible assignments to find the one which maximises the objective value. We do this using backtracking search interleaved with constraint propagation, using the min-dom and min-value variable and value selection heuristics. The model is implemented in and solved with CP-Inside [13]. We find the optimal in, on average, 250 milliseconds.

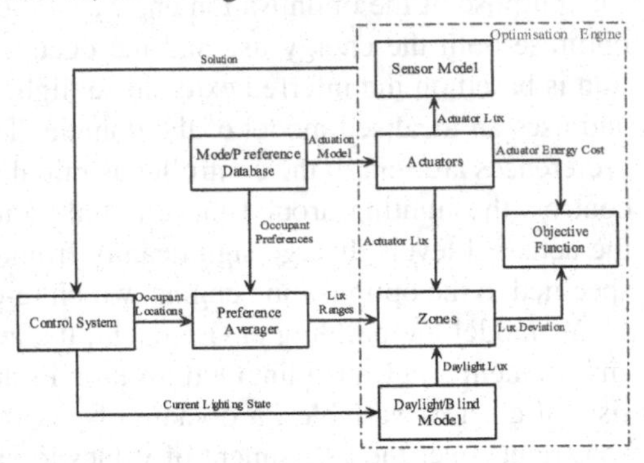

Fig. 5 Opimisation Process

Note that the optimisation model is not an accurate control model. It is a simplified model of actuators and of their effect on the environment, and it assumes the actuation is perfect, the propagation of light is uniform, and that the sensors are perfect. It also does not account for small variations in the external light. Instead, from the optimal solution we derive the intended lux levels for each zone, and pass them to the controller as the target lux levels. The values for the lights are passed as initial settings, and the controller then controls the actuated lights in order to achieve the target lux levels. The controller thus handles the inaccuracies in the actuation and sensing, and any small variations in external light. When the controller changes the set points of the lights to such an extent that a light is 2 or more set points away from its initial setting, we assume that the external light has changed, and so we re-state the problem to the optimiser, which computes a new set of intended lux levels and a new initial settings for the lights. Similarly, when we detect that an occupant has left, arrived, or changed position, we again pass this new problem to the optimiser.

4 Simulation Results

The Charon model described earlier and its environment have been simulated using the Charon simulation tool-set. In this section we provide the simulation results and evaluate the power saving comparing to a typical control technique used in building automation.

4.1 *Lighting Control*

In order to observe the control- and the environment behaviour, Fig. 6(e) shows the external light coming to the controlled area and how it is affected by the blinding occlusion changes of Fig. 6(f) (5:06 - 5:24pm and 7:24-7:54 pm) and the clouds move (11:36am-4:54 pm). However, Fig. 6(c) shows the person movements among the zones and their effect on the optimal light that calculated by the optimisation engine for each zone, as depicted in Fig. 6(b).

To evaluate the control strategy, we consider 2 zones: the first one (Zone 1) is highly affected by the daylight whereas the other one (Zone 6) is less affected. Regarding internal light in Zone 1 (shown in Fig .6(a)), when there is no presence in zone 1, the internal light is affected only by the external light and the interferences from the neighbour zones. However when a person comes inside, internal light 1 gets actuated to reach the optimal light decided by the optimiser. Zone 6 follows the same routine, but with reduced daylight influence; for example at 9:42 pm, when zone 6 is unoccupied, internal light 6 equals to 82 Lux which mainly comes from the light interference.

4.2 *Energy Saving*

In order to evaluate the potential energy efficiency of the proposed control strategy, we have considered a typical control strategy used in building automation as base-line for comparison. The base-line model uses Passive Infrared (PIR) sensor for presence detection in order to switch on the artificial light to a predefined luminance level. In our case, we consider 350 Lux as the optimal light in the entire zone since it is almost the average optimal intensity requested as in Fig. 6(b). Due to the fact that energy consumption has a linear relation with the consumed Lux over time, we have compared the consumed Lux in each case as shown in Fig. 6(d). The results show that the proposed control strategy reduces energy consumption by 42% in comparison to the baseline model.

5 Conclusion

In this article we have proposed a hybrid/muti-agent model for an intelligent automated lighting system. The control strategy maintains user comfort through systematically tracking occupants in each zone in order to integrate their preferences. The control system incorporates a constraint-based optimiser that computes the optimal setting, thereby optimising energy usage while providing adequate user comfort. The simulation results shows that the proposed controller saves around 42% of the energy consumption compared to a standard baseline control strategy.

As future work, we intend to deploy a demonstration of the developed system in the Environmental Research Institute (ERI) building, which is the ITOBO Living Laboratory [14].

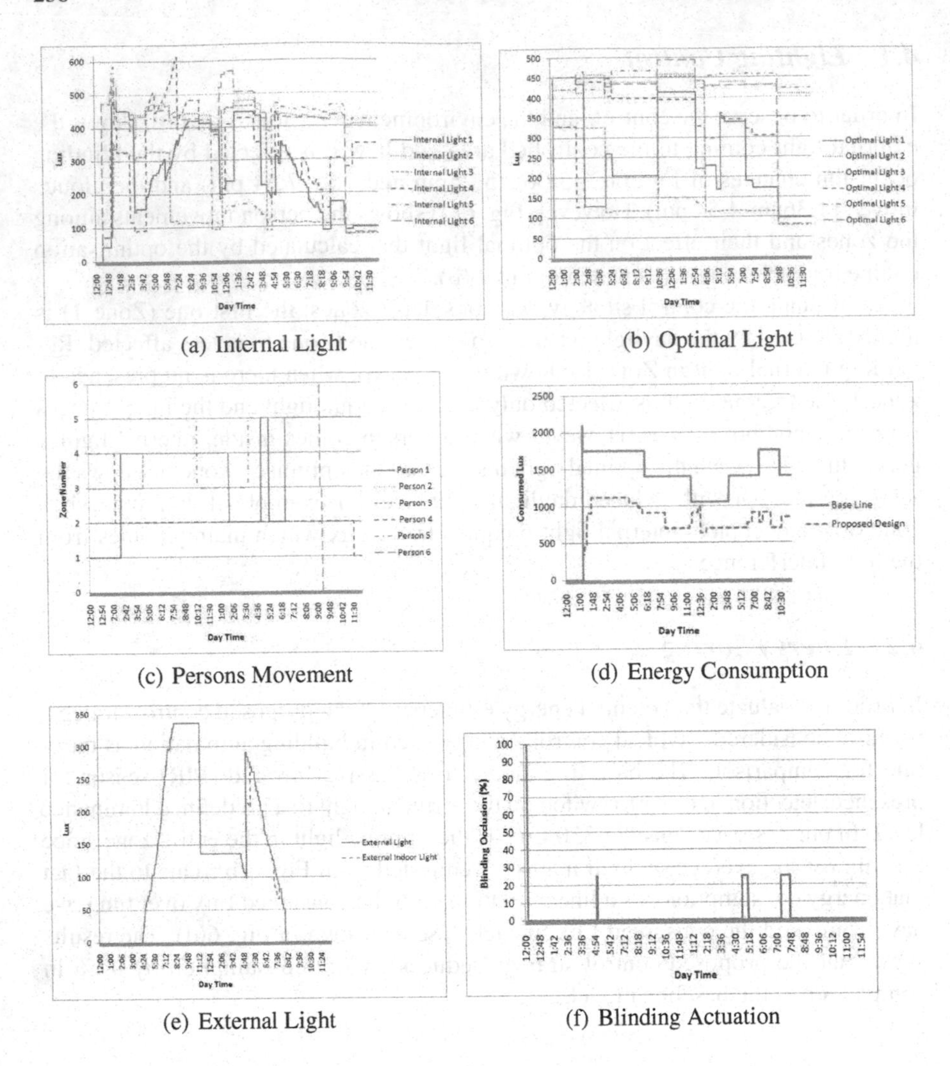

(a) Internal Light (b) Optimal Light

(c) Persons Movement (d) Energy Consumption

(e) External Light (f) Blinding Actuation

Fig. 6 Model Simulation Results

Acknowledgements. This work was funded by Science Foundation Ireland (SFI) grant 06-SRC-I1091.

References

1. The Interlaboratory Working Group on Energy-Efficient and Clean-Energy, Scenarios for a Clean Energy Future: Interlaboratory Working Group on Energy-Efficient and Clean-Energy Technologies (2000)
2. Labinaz, G., Bayoumi, M.M., Rudie, K.: Modeling and Control of Hybrid Systems: A Survey. In: Proc. of the 13th Triennal World Congress, San Francisco, USA (1996)

3. Charon: Modular Specification of Hybrid Systems, http://rtg.cis.upenn.edu/mobies/charon/
4. Gössler, G., Sifakis, J.: Composition for Component-Based Modeling. In: de Boer, F.S., Bonsangue, M.M., Graf, S., de Roever, W.-P. (eds.) FMCO 2002. LNCS, vol. 2852, pp. 443–466. Springer, Heidelberg (2003)
5. Keppens, J., Shen, Q.: On compositional modeling. In: The Knowledge Engineering Review, pp. 157–200 (2001)
6. Booch, G., Rumbaugh, J., Jacobson, I.: The Unified Modeling Language User Guide. Addison Wesley, Reading (1998)
7. Mady, A., Boubekeur, M., Provan, G.: Compositional Model-Driven Design of Embedded Code for Energy-Efficient Buildings. In: 7th IEEE International Conference on Industrial Informatics (INDIN 2009), Cardiff, UK (2009)
8. Romer, K., Kasten, O., Mattern, F.: Middleware Challenges for Wireless Sensor Networks. In: ACM SIGMOBILE Mobile Computing and Communications Review (2002)
9. Kolokotsa, D., Pouliezos, A., Stavrakakis, G., Lazos, C.: Predictive control techniques for energy and indoor environmental quality management in buildings. In: Building and Environment (2008)
10. Marshall, A.H., McClean, S.I.: Using Coxian Phase-Type Distributions to Identify Patient Characteristics for Duration of Stay in Hospital. In: Health Care Management Science (2004)
11. Henzinger, T.: The theory of hybrid automata. In: Proc. 11th Annual IEEE Symposium on Logic in Computer Science (LICS 1996), pp. 278–292 (1996)
12. Dechter, R.: Constraint Processing. Morgan Kaufmann Publishers, San Francisco (2003)
13. Feldman, J., Freuder, E., Little, J.: CP-INSIDE: Embedding Constraint-Based Decision Engines in Business Applications. In: Integration of AI and OR Techniques in Constraint Programming for Combinatorial Optimization Problems (2009)
14. Environmental Research Institute, http://www.ucc.ie/en/ERI/

Author Index